T0213650

Lecture Notes in Computer Science   10443

Commenced Publication in 1973
Founding and Former Series Editors:
Gerhard Goos, Juris Hartmanis, and Jan van Leeuwen

More information about this series at http://www.springer.com/series/7409

Miroslav Bursa · Andreas Holzinger
M. Elena Renda · Sami Khuri (Eds.)

Information Technology in Bio- and Medical Informatics

8th International Conference, ITBAM 2017
Lyon, France, August 28–31, 2017
Proceedings

 Springer

Editors
Miroslav Bursa
Czech Technical University
Prague
Czech Republic

Andreas Holzinger
Medical University Graz
Graz
Austria

M. Elena Renda
Institute of Informatics and Telematics
Pisa
Italy

Sami Khuri
San José State University
San José, CA
USA

ISSN 0302-9743   ISSN 1611-3349 (electronic)
Lecture Notes in Computer Science
ISBN 978-3-319-64264-2   ISBN 978-3-319-64265-9 (eBook)
DOI 10.1007/978-3-319-64265-9

Library of Congress Control Number: 2017947507

LNCS Sublibrary: SL3 – Information Systems and Applications, incl. Internet/Web, and HCI

Printed on acid-free paper

This Springer imprint is published by Springer Nature
The registered company is Springer International Publishing AG
The registered company address is: Gewerbestrasse 11, 6330 Cham, Switzerland

Preface

Biomedical engineering and medical informatics represent challenging and rapidly growing areas. Applications of information technology in these areas are of paramount importance. Building on the success of ITBAM 2010, ITBAM 2011, ITBAM 2012, ITBAM 2013, ITBAM 2014, ITBAM 2015, and ITBAM 2016, the aim of the 8th International Conference on Information Technology in Bio- and Medical Informatics (ITBAM) conference was to continue bringing together scientists, researchers, and practitioners from different disciplines, namely, from mathematics, computer science, bioinformatics, biomedical engineering, medicine, biology, and different fields of life sciences, to present and discuss their research results in bioinformatics and medical informatics.

We hope that ITBAM will continue to serve as a platform for fruitful discussions between all attendees, where participants can exchange their recent results, identify future directions and challenges, initiate possible collaborative research and develop common languages for solving problems in the realm of biomedical engineering, bioinformatics, and medical informatics. The importance of computer-aided diagnosis and therapy continues to draw attention worldwide and has laid the foundations for modern medicine with excellent potential for promising applications in a variety of fields, such as telemedicine, Web-based health care, analysis of genetic information, and personalized medicine.

Following a thorough peer-review process, we finally selected three long papers for oral presentation and six short papers for poster session for the eighth annual ITBAM conference (from a total of 15 contributions). The Organizing Committee would like to thank the reviewers for their excellent job. The articles can be found in the proceedings. The papers show how broad the spectrum of topics in applications of information technology to biomedical engineering and medical informatics is.

The editors would like to thank all the participants for their high-quality contributions and Springer for publishing the proceedings of this conference. Once again, our special thanks go to Gabriela Wagner for her hard work on various aspects of this event.

June 2017
M. Elena Renda
Miroslav Bursa
Andreas Holzinger
Sami Khuri

Organization

General Chair

Christian Böhm — University of Munich, Germany

Program Committee Co-chairs

Miroslav Bursa — Czech Technical University, Czech Republic
Andreas Holzinger — Medical University Graz, Austria
Sami Khuri — San José State University, USA
M. Elena Renda — IIT, CNR, Pisa, Italy

Program Committee

Akutsu, Tatsuya	Kyoto University, Japan, Japan
Albrecht, Andreas	Middlesex University London, UK
Baumann, Peter	Jacobs University Bremen, Germany
Bursa, Miroslav	Czech Technical University in Prague, Czech Republic
Böhm, Christian	University of Munich, Germany
Casadio, Rita	University of Bologna, Italy
Casillas, Sònia	Universitat Autònoma de Barcelona, Spain
Ehrich, Hans-Dieter	Technical University of Braunschweig, Germany
Friedrich, Christoph M.	University of Applied Sciences Dortmund, Germany
Havlik, Jan	Czech Technical University in Prague, Czech Republic
Heun, Volker	Ludwig-Maximilians-Universität München, Germany
Holzinger, Andreas	Technical University Graz and Medical Informatics, Medical University Graz, Austria
Ismailova, Larisa	NRNU MEPhI, Moscow, Russian Federation
Kerr, Alastair	University of Edinburgh, UK
Khuri, Sami	San Jose State University, USA
Kuzilek, Jakub	Czech Technical University, Czech Republic
Lhotska, Lenka	Czech Technical University, Czech Republic
Marshall, Roger	Plymouth State University, USA
Masciari, Elio	ICAR-CNR, Università della Calabria, Italy
Pisanti, Nadia	University of Pisa, Italy
Pizzi, Cinzia	University of Padova, Italy
Pizzuti, Clara	Institute for High Performance Computing and Networking, National Research Council, Italy
Renda, Maria Elena	CNR-IIT, Italy
Rovetta, Stefano	University of Genoa, Italy
Santana, Roberto	University of the Basque Country, Spain
Seker, Huseyin	University of Northumbria at Newcastle, UK

Spilka, Jiri	Czech Technical University in Prague, Czech Republic
Steinhofel, Kathleen	King's College London, UK
Zhang, Songmao	Chinese Academy of Sciences, China
Zhu, Qiang	University of Michigan, USA

Additional Reviewer

Corrado Pecori	Università e-Campus, Italy

Contents

Editorial

IT in Biology & Medical Informatics: On the Challenge of Understanding the Data Ecosystem

Andreas Holzinger[1]([⊠]), Miroslav Bursa[2], Sami Khuri[3], and M. Elena Renda[4]

[1] Holzinger Group, HCI-KDD, Institute for Medical Informatics and Statistics,
Medical University Graz, Graz, Austria
andreas.holzinger@medunigraz.at
[2] Czech Institute of Informatics, Robotics and Cybernetics,
Czech Technical University in Prague, Prague, Czech Republic
miroslav.bursa@cvut.cz
[3] Department of Computer Science, San Jose State University, San Jose, CA, USA
sami.khuri@sjsu.edu
[4] Istituto di Informatica e Telematica - CNR, Pisa, Italy
elena.renda@iit.cnr.it

Abstract. Data intensive disciplines, such as life sciences and medicine, are promoting vivid research activities in the area of data science. Modern technologies, such as high-throughput mass-spectrometry and sequencing, micro-arrays, high-resolution imaging, etc., produce enormous and continuously increasing amounts of data. Huge public databases provide access to aggregated and consolidated data on genome and protein sequences, biological pathways, diseases, anatomy atlases, and scientific literature. There has never been before more potentially available data to study biomedical systems, ranging from single cells to complete organisms. However, it is a non-trivial task to transform the vast amount of biomedical data into actionable, useful and usable information, triggering scientific progress and supporting patient management.

Keywords: Biomedical informatics · Data science · Data ecosystem

1 Life Science and Medicine as Data Science

The fact that life sciences and medicine are turning into data sciences is well accepted and is widely adopted ranging from the use of classic hospital data for secondary research [1] to the use of data from wearable devices for health, wellness and wellbeing [2]. Data intensive medicine has a lot of promises, e.g. getting closer to personalized medicine, where the goal is not only to support evidence-based decision-making, but to tailor decisions, practices and therapies to the individual patient [3]. As this is a grand goal for the future, posing a lot of open questions and challenges, stratified medicine is state-of-the-art, which aims to select at least the optimal therapy for groups of patients who share common biological characteristics [4]. Machine Learning methods can be of help here,

© Springer International Publishing AG 2017
M. Bursa et al. (Eds.): ITBAM 2017, LNCS 10443, pp. 3–7, 2017.
DOI: 10.1007/978-3-319-64265-9_1

for example *causal inference trees (CIT)* and aggregated grouping, seeking strategies for deploying stratified approaches [5]. We provide this example because this demonstrates a huge and increasing challenge: it is not the amount of the data that is the problem, but the increasing heterogeneity of the data! This increasing amount of *heterogeneous data sets,* in particular "-omics" data (genomics, proteomics, metabolomics, etc.) [6] make traditional data analysis methods problematic. Complex chronic diseases are also very heterogeneous across individual patients – which call for integrative analysis approaches [7].

Interestingly, many large data sets are indeed large collections of small data sets – but complex data. This is particularly the case in personalized medicine where there might be a large amount of data, but there is still a relatively small amount of data for each patient available [8].

In order to customize predictions for each individual, it is necessary to build models for each patient along with the inherent uncertainties, and to couple these models together in a hierarchy so that information can be "borrowed" from other similar patients [9]. This is called *model personalization,* and is naturally implemented by using hierarchical Bayesian approaches including e.g. hierarchical Dirichlet processes [10] or Bayesian multi-task learning [11].

2   Challenge: Understanding the Data Ecosystem

To help solving such problems, the first and most important step is to gain a clear understanding of the data ecosystem. This includes issues of Data Collection, *Data Integration, Data Fusion, Data Preprocessing,* and *Data Visualization.*

On one hand we have biomedical *research* data, e.g. clinical trial data, -omics data, e.g. from genomic sequencing technologies (Next Generation Sequencing, NGS etc.), microarrays, transcriptomic technologies, etc., which all play important roles for biomarker discovery and drug design [12]. In this area the vast amount of large and complex data sets in high dimensional spaces makes manual analysis practically impossible and calls for the application of automatic machine learning applications for the analysis of e.g. genome sequencing data sets, including the annotation of sequence elements and epigenetic, proteomic or metabolomic data [13].

On the other hand, we have large amounts of clinical data, e.g. patient records, medical documentations, terminologies (e.g. ICD-10, SNOMED-CT, OMMIM, SNOMED), medical study and survey data, laboratory data, e.g. clinical and physiological parameters, and a large pool of signal data ECG, EOG, EEG etc., and image data (from radiology, confocal laser scanning, etc.). Here we sometimes are confronted with little data, or rare events, where automatic approaches suffer of insufficient training samples. Here interactive machine learning may be of help, particularly with a doctor-in-the-loop, e.g. in subspace clustering, k-Anonymization, protein folding and protein design [14].

However, data can also be collected from unexpected sources, like cities' wastewater systems, pursuing the idea that cities can make use of their waste water system to do near real-time urban epidemiology: since a broad array of human activity is reflected in a city's wastewater, tapping into this vast reservoir of information can help monitor urban health patterns, shaping more inclusive public health strategies, and pushing the boundaries of urban epidemiology (see e.g. http://underworlds.mit.edu).

In times of financial cuts, health business data becomes important, e.g. utilization, management data, accounting etc. including prediction and forecasting. Moreover, data collected on a private basis, e.g. sampled from wellness devices, sport gadgets, ambient assistant living environments, etc., may demonstrate the usefulness of IT for ensuring a healthy society [15].

Data preprocessing is often underestimated but can be of vital importance, particularly as pre-processing step for feeding the data into machine learning pipelines, e.g. big data is often needed for learning approaches, but need to be combined from different heterogeneous data sources; ignoring the difficulties can result in biased results, modeling of artifacts and even misleading inference. [16].

Data integration is a hot topic generally and in health informatics specifically and solutions can bridge the gap between clinical and biomedical research [17]. This is becoming even more important due to the increasing amounts of heterogeneous, complex patient related data sets, resulting from various sources including picture archiving and communication systems (PACS) and radiological information systems (RIS), hospital information systems (HIS), laboratory information systems (LIS), physiological and clinical data repositories, and all sorts of -omics data from laboratories, using biological samples from Biobanks (which can be of utmost importance for the study of diseases [18]). Such Biobanks include large collections of all sort of tissue, body fluids, etc., but also DNA sequence data, proteomic and metabolic data; resulting from sophisticated high-throughput analytical technologies. Along with classical patient records, containing large amounts of unstructured and semi-structured information, integration efforts incorporate enormous problems, but at the same time offers new possibilities for translational research - which opens avenues e.g. for cancer research.

While data integration is the combination of data from different sources and providing users with a unified view on these data (e.g. combining research results from different bioinformatics repositories), *data fusion* is matching various data sets which represent one and the same object into a single, consistent, and clean representation [19]. In health informatics these unified views are important in high-dimensional spaces, e.g. for integrating heterogeneous descriptions of the same set of genes [20]. The main expectation is that such fused data is more informative than the original inputs, hence particularly valuable for future research.

Data visualization has the main goal of communicating quantitative information in a clear and effective way through graphical means, and could greatly enhance our ability to make sense of data mining results. It has been proven that visual representation of -omics data could be highly beneficial since it helps to view a large amount of information at a time, and it also allows to easily identify patterns and trends from a massive volume of data [21].

Capturing all information describing a biological system is the implicit objective of all -omics methods, however, genomics, transcriptomics, proteomics, metabolomics, etc. need to be combined to approach this goal: valuable information can be obtained using various analytical techniques such as nuclear magnetic resonance, liquid chromatography, or gas chromatography coupled to mass spectrometry. Each method has inherent advantages and disadvantages, but are complementary in terms of biological information, consequently combining multiple data sets, provided by different analytical platforms is of utmost importance. For each platform, the relevant information is extracted in the first step. The obtained latent variables are then fused and further analyzed. The influence of the original variables is then calculated back and interpreted. There is plenty of open future research to include all possible sources of information [22], which calls for machine learning approaches [23].

3   Conclusion

There are huge challenges and a lot of open problems in data science generally and in data integration, data fusion, data preprocessing specifically. A hard task is to map the results gained in arbitrarily high dimensional spaces, down into the lower dimensions to make them accessible to a human end user. In essence, successful application of sophisticated information technology in Biology and Medical Informatics requires a clear and deep understanding of the data ecosystem!

Acknowledgments. We are grateful for the great support of the ITBAM program committee, and particularly for the excellent work of Gabriela Wagner.

References

1. Andersen, T.F., Madsen, M., Jorgensen, J., Mellemkjaer, L., Olsen, J.H.: The Danish national hospital register - a valuable source of data for modern health sciences. Dan. Med. Bull. **46**, 263–268 (1999)
2. Sathyanarayana, A., Srivastava, J., Fernandez-Luque, L.: The science of sweet dreams: predicting sleep efficiency from wearable device data. Computer **50**, 30–38 (2017)
3. Holzinger, A.: Trends in interactive knowledge discovery for personalized medicine: cognitive science meets machine learning. IEEE Intell. Inform. Bull. **15**, 6–14 (2014)
4. Trusheim, M.R., Berndt, E.R., Douglas, F.L.: Stratified medicine: strategic and economic implications of combining drugs and clinical biomarkers. Nat. Rev. Drug Discovery **6**, 287–293 (2007)
5. Su, X., Kang, J., Fan, J., Levine, R.A., Yan, X.: Facilitating score and causal inference trees for large observational studies. J. Mach. Learn. Res. **13**, 2955–2994 (2012)
6. Huppertz, B., Holzinger, A.: Biobanks – a source of large biological data sets: open problems and future challenges. In: Holzinger, A., Jurisica, I. (eds.) Interactive Knowledge Discovery and Data Mining in Biomedical Informatics. LNCS, vol. 8401, pp. 317–330. Springer, Heidelberg (2014). doi:10.1007/978-3-662-43968-5_18

7. Schulam, P., Saria, S.: Integrative analysis using coupled latent variable models for individualizing prognoses. J. Mach. Learn. Res. **17**, 1–35 (2016)

8. Rost, B., Radivojac, P., Bromberg, Y.: Protein function in precision medicine: deep understanding with machine learning. FEBS Lett. **590**, 2327–2341 (2016)

9. Ghahramani, Z.: Bayesian non-parametrics and the probabilistic approach to modelling. Philos. Trans. R. Soc. A: Math. Phys. Eng. Sci. **371**, 20110553 (2013)

10. Teh, Y.W., Jordan, M.I., Beal, M.J., Blei, D.M.: Hierarchical Dirichlet processes. J. Am. Stat. Assoc. **101**, 1566–1581 (2006)

11. Houlsby, N., Huszar, F., Ghahramani, Z., Hernández-lobato, J.M.: Collaborative Gaussian processes for preference learning. In: Pereira, F., Burges, C., Bottou, L., Weinberger, K. (eds.) Advances in Neural Information Processing Systems (NIPS 2012), pp. 2096–2104 (2012)

12. McDermott, J.E., Wang, J., Mitchell, H., Webb-Robertson, B.J., Hafen, R., Ramey, J., Rodland, K.D.: Challenges in biomarker discovery: combining expert insights with statistical analysis of complex omics data. Expert Opin. Med. Diagn. **7**, 37–51 (2013)

13. Libbrecht, M.W., Noble, W.S.: Machine learning applications in genetics and genomics. Nat. Rev. Genet. **16**, 321–332 (2015)

14. Holzinger, A.: Machine learning for health informatics. In: Holzinger, A. (ed.) Machine Learning for Health Informatics: State-of-the-Art and Future Challenges. LNCS, vol. 9605, pp. 1–24. Springer, Cham (2016). doi:10.1007/978-3-319-50478-0_1

15. Varshney, U., Chang, C.K.: Smart health and well-being. Computer **49**, 11–13 (2016)

16. Tang, L., Song, P.X.: Fused lasso approach in regression coefficients clustering - learning parameter heterogeneity in data integration. J. Mach. Learn. Res. **17**, 1–23 (2016)

17. Jeanquartier, F., Jean-Quartier, C., Schreck, T., Cemernek, D., Holzinger, A.: Integrating open data on cancer in support to tumor growth analysis. In: Renda, M.E., Bursa, M., Holzinger, A., Khuri, S. (eds.) ITBAM 2016. LNCS, vol. 9832, pp. 49–66. Springer, Cham (2016). doi:10.1007/978-3-319-43949-5_4

18. Gottweis, H., Zatloukal, K.: Biobank governance: trends and perspectives. Pathobiology **74**, 206–211 (2007)

19. Bleiholder, J., Naumann, F.: Data fusion. ACM Comput. Surv. (CSUR) **41**, 1–41 (2008)

20. Lafon, S., Keller, Y., Coifman, R.R.: Data fusion and multicue data matching by diffusion maps. IEEE Trans. Pattern Anal. Mach. Intell. **28**, 1784–1797 (2006)

21. Pellegrini, M., Renda, M.E., Vecchio, A.: Tandem repeats discovery service (TReaDS) applied to finding novel cis-acting factors in repeat expansion diseases. BMC Bioinformatics **13**, S3 (2012)

22. Blanchet, L., Smolinska, A.: Data fusion in metabolomics and proteomics for biomarker discovery. In: Jung, K. (ed.) Statistical Analysis in Proteomics, pp. 209–223. Springer, New York (2016)

23. Holzinger, A.: Introduction to Machine Learning and Knowledge Extraction (MAKE). Mach. Learn. Knowl. Extr. **1**(1), 1–20 (2017). doi:10.3390/make1010001

General Track

A Hybrid Feature Selection Method to Classification and Its Application in Hypertension Diagnosis

Hyun Woo Park, Dingkun Li, Yongjun Piao, and Keun Ho Ryu[✉]

Database and Bioinformatics Laboratory, School of Electrical and Computer Engineering,
Chungbuk National University, Cheongju, South Korea
{hwpark,jerryli,pyz,khryu}@dblab.chungbuk.ac.kr

Abstract. Recently, various studies have shown that meaningful knowledge can be discovered by applying data mining techniques in medical applications, i.e., decision support systems for disease diagnosis. However, there are still several computational challenges due to the high-dimensionality of medical data. Feature selection is an essential pre-processing procedure in data mining to identify relevant feature subset for classification. In this study, we proposed a hybrid feature selection mechanism by combining symmetrical uncertainty and Bayesian network. As a case study, we applied our proposed method to the hypertension diagnosis problem. The results showed that our method can improve the classification performance and outperformed existing feature selection techniques.

Keywords: Classification · Feature selection · Hypertension · KNHANES · Data mining

1 Introduction

In recent years, the proportion of elderly people (over 65 years old) is increasing in Korea. The proportion of elderly people population will increase up to 30% by 2030 according to a recent report by Korea National Statistical Office [1]. Therefore, the number of chronic disease patients is increasing as well according to elderly people increasing. Approximately 80% of men and women 70 years of age have at least one of chronic disease such as hypertension, heart disease and so on [2]. Furthermore, chronic disease is continuous growth of health care cost. According to the report, health care costs for Koreans over age 65, reached 15.4 trillion Korean won in 2011 [3]. In general, chronic disease cannot be prevented by vaccines or cured by medication. Most of the common chronic disease caused by dietary, lifestyle (smoking, drinking), and many other factors. The hypertension is also one of the chronic disease type, the prevalence of hypertension is increasing according to Korean National Health and Nutrient Examination Survey report in 2011 [4]. The hypertension is a major risk factor for heart disease, and many other complications and these complication leading to death. Therefore, the prevention of hypertension has become a major issue in the world.

Most previous studies used the statistical method such as chi-square, and logistic regression for finding risk factors pertaining to chronic diseases [5–7]. Bae et al. [5] investigated the association between hypertension and prevalence of low back pain and

© Springer International Publishing AG 2017
M. Bursa et al. (Eds.): ITBAM 2017, LNCS 10443, pp. 11–19, 2017.
DOI: 10.1007/978-3-319-64265-9_2

osteoarthritis in Koreans using Chi-square and logistic regression. Song et al. [6] examined the associations of total carbohydrate intake, dietary glycaemic load (DGL) and white rice intake with metabolic syndrome risk factors by gender in Korean adolescents using multivariate linear regression. Ha et al. [7] examined whether cardiovascular disease or its risk factors were associated with chronic low back pain (cLBP) in Koreans using logistic regression. The regression models allow for the testing of statistical interactions among independent variables and significance difference in the effects of one or more independent variables.

Recently, various studies [8–13] have shown that it is possible to apply data mining techniques in medical applications [14–17], i.e., decision support systems for disease diagnosis. However, medical data generally contains several irrelevant and redundant features regarding classification target. Those features may lead to low performance and high computational complexity of disease diagnosis. Moreover, most classification methods assume that all features have uniform importance degree during classification [18]. Thus, dimension reduction techniques that discover a reduced set of features are needed to achieve better classification result.

In this study, we proposed a hybrid feature selection method to improve the robustness and the accuracy of the hypertension diagnosis. Symmetrical uncertainty was used to preliminarily remove irrelevant features as a filter and correlation between two features is compared and the lower symmetrical uncertainty is removed. Machine learning algorithms with backward search were used as the wrapper part. The results showed that the proposed method yielded good performance and outperformed other feature selection approaches. The results, although preliminary, are expected to support medical decision making, and consequently reduce the expenditure in medical care.

The reminder of the paper is organized as follows. In Sect. 2, we describe dataset and present hybrid feature selection, classification method. In Sect. 3, shows the framework of experiment and results. The conclusion and future are presented in Sect. 4.

2 Materials and Methods

2.1 Data

In this study, we conducted Korea National Health and Nutrient Examination Survey (KNHNAES). This data set is a national project and it is consisting of four parts. The first part of the survey recorded the socio-demographic characteristics include age, gender, income, education level and so on. The second part of the survey recorded the history of a disease and a third part is recorded health medical examination such as blood pressure (systolic, diastolic). The last part of the survey recorded life pattern and nutritional intake. We conducted KNHANES data from 2007 to 2014. This data set contains lots of missing values and outliers. This data may lead to poor performance, we eliminated this data for generating target population. The basic characteristics of the target population are shows in Table 1.

Table 1. Basic characteristics of target population

	Control (n = 2,938)	Hypertension (n = 2,700)
Age (yr)	45.51 ± 15.46	65.20 ± 10.93
Sex		
Male	1,066	1,151
Female	1,872	1,549
Education		
High school	1,800	2,427
University	1,138	273
Smoking		
Yes	503	574
No	499	1,651
Quick	1,936	475
SBP (mmHg)	112.84 ± 15.00	131.63 ± 16.21
DBP (mmHg)	73.56 ± 9.68	80.53 ± 9.02
BMI (kg/m^2)	22.91 ± 3.11	24.70 ± 3.18
Waist Circumference (cm)	77.79 ± 9.17	85.28 ± 8.93

2.2 Feature Selection

Feature selection is an essential pre-processing procedure in data mining for identifying relevant subset for classification. The high dimensionality of the data may cause a various problem such as increasing the complexity and reducing the accuracy, i.e. curse of dimensionality. The goal of feature selection is to provide faster construction of prediction models with a better performance [8]. Feature selection approaches can be broadly grouped into three categories: filter, wrapper, and hybrid [19]. The main difference of filter and wrapper method is in whether they adopt a machine learning algorithm to guide the feature selection or not. In general, filters employ independent evaluation measures thus are fast but can generate the local-optimal result. In contrast, wrapper methods adopt a searching algorithm to iteratively generate several subsets, evaluate them based on the classification algorithm, and finally choose the subset with best classification performance. Wrappers usually can produce better results than filters but they are computationally expensive. Hybrid methods combined the advantages of filter and wrapper techniques to achieve better learning performance with a similar computational cost of filters [20]. A feature selection procedure can usually be divided into two steps: subset generation and subset evaluation. The most important this process is determined to search strategy and the starting point. Sequential search method, such as Sequential Forward Search (SFS) and Sequential Backward Search (SBS). SFS method start with an empty candidate set and add features until the addition of features does not decrease the criterion. SBS method is removed from a full candidate set until the removal of further features increases the criterion. We illustrate the filter and wrapper approach in Fig. 1.

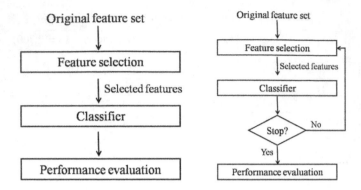

Fig. 1. (a) Filter approach and (b) wrapper approach

In this study, we proposed a hybrid feature selection method to improve the robustness and the accuracy of the hypertension diagnosis. Here, we describe our proposed hybrid feature selection method. The overall framework of generating optimal feature subset is illustrated in Fig. 2. In the proposed method, symmetrical uncertainty was used to preliminarily remove irrelevant features and remove redundant features used Pearson correlation as a filter. Bayesian network [21] with backward search [22] adopted as the wrapper part.

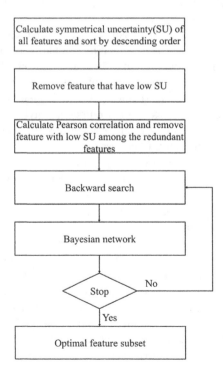

Fig. 2. Framework of proposed feature selection method

For each feature in the original space, the symmetrical uncertainty value is calculated and the features that have smaller symmetrical uncertainty value than the predefined threshold are removed. Then the remaining features are sorted in the descending order of their symmetrical uncertainty value. We compared the correlation between two features and remove the features with low gain ratio among the two features whose correlation is higher than the predefined threshold. Afterward, starting from the whole features generated from the filter part, removing one feature at one time, the subset is evaluated by the Bayesian network until the best feature subset that has the highest accuracy is selected.

The Information gain (IG) is the decrease in the entropy of Y for given information on Y provided by X and is calculated using (3), which is based on (1) and (2). Then the measure of symmetrical uncertainty (SU) is calculated using (4), which can compensate for the problem of a biased IG and normalize its value from 0 to 1.

$$H(Y) = -\sum_{y \in Y} p(y) \log_2 p(y) \tag{1}$$

$$H(Y|X) = -\sum_{x \in X} p(x) \sum_{y \in Y} p(y|x) \log_2 p(y|x) \tag{2}$$

$$gain = H(Y) + H(X) - H(X|Y) \tag{3}$$

$$SU = 2.0 \times \left[\frac{gain}{H(Y) + H(X)} \right] \tag{4}$$

Where H(X) is the entropy of feature X, H(C) indicates the entropy of the class, and H(X|C) is the entropy of x after observing class C.

3 Experiment and Results

The framework of the experiment is shown in Fig. 3. Firstly, we need to generate of the target population based on KNHANES dataset. In the next step, we remove missing value and outliers based interquartile ranges and we used proposed feature selection method to remove irrelevant and redundant features. Then, we compared the performance of classification with several commonly used feature selection methods such as Information gain, Gain ratio, ReliefF.

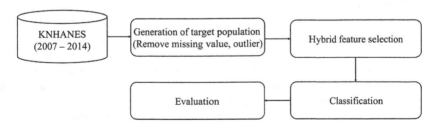

Fig. 3. Framework of experiment

3.1 Generation of Target Population

In this study, we conducted KNHANES from 2007 to 2014. This data contains 65,973 individuals include 9,383 hypertensions patients' data. We eliminated a considerable number of hypertension patients with following exclusion criteria to avoid data bias. The target population data contains 5,698 samples (2,938 hypertensions and 2,700 controls) samples. We describe the procedure of generation of the target population in Fig. 4.

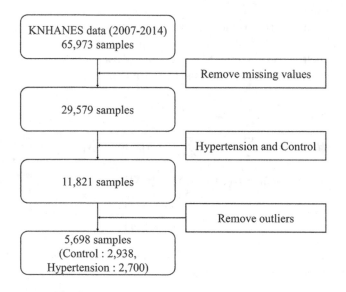

Fig. 4. Procedure of generating target population

3.2 Hybrid Feature Selection

First of all, we eliminated irrelevant features whose SU values are less than 0.01, and sort in descending order. Afterward, we remove redundant features using Pearson correlation between two features. We compared the correlation with the property with highest SU value and remove the variables with low SU value among the features with a correlation greater than 0.5. In the next step we used Bayesian network with SBS search, then we find optimal feature subset.

3.3 Bayesian Network

A Bayesian network is based on probability theory and graphical model. The graph consists of one or more nodes and edges. Each node is the graph represents a random variable, and each arc between every pair of variables. Although the sample size is not large enough for estimating the predictive performance of each model, Bayesian network in general has good prediction abilities.

3.4   Experimental Results

To achieve reliable results, we conducted 10-fold cross-validation during all the experiments. In 10-fold cross-validation, 9 parts were used to train the model and the remaining one was used to test the model. Moreover, we adopted several evaluation measures such as F-measure, sensitivity, specificity, and area under ROC (AUC) to the performance of classification. Table 2 shows the performance of our proposed method. The results were found to be 0.923, 0.923, 0.923, and 0.975 for F-measure, sensitivity, specificity, and AUC, respectively. In Table 2, hypertension indicates the patients with hypertension disease and control indicates those without hypertension and FS refers to apply proposed hybrid feature selection and Non-FS indicates without feature selection. Figure 5. Shows the classification accuracy of 4 feature selection methods. From the Fig. 5. We can easily see that our proposed method outperforms existing feature selection approaches.

Table 2. Classification results of the proposed method

		Hypertension	Control	Average
F-Measure	Non-FS	0.879	0.885	0.882
	Proposed FS	**0.921**	**0.926**	**0.923**
Sensitivity	Non-FS	0.896	0.869	0.882
	Proposed FS	**0.928**	**0.919**	**0.923**
Specificity	Non-FS	0.869	0.896	0.882
	Proposed FS	**0.919**	**0.928**	**0.923**
AUC	Non-FS	0.955	0.955	0.955
	Proposed FS	**0.975**	**0.975**	**0.975**

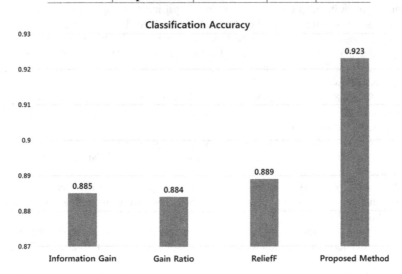

Fig. 5. Performance comparison of the proposed method with other feature selection techniques

4   Conclusion

In this study, we proposed a hybrid feature selection technique based on symmetrical uncertainty and Pearson correlation as a filter approach and Bayesian network as a wrapper approaches for accurately classify hypertensions. To validate the proposed method, we conducted KNHANES 2007–2014 data. We conducted several experiments and compared the performance with existing feature selection approaches. The results showed the proposed method had good performance and outperformed the other feature selection methods.

The hypertension is a major risk factor for heart disease, and many other complications and these complication leading to death. In the future work, we will analyze hypertension and other complications disease such as heart disease, stroke.

Acknowledgment. This research was supported by the MSIP (Ministry of Science, ICT and Future Planning), Korea, under the ITRC (Information Technology Research Center) support program (IITP-2017-2013-0-00881) supervised by the IITP (Institute for Information & communication Technology Promotion) and Basic Science Research Program through the National Research Foundation of Korea (NRF) funded by the Ministry of Science, ICT & Future Planning (No. 2017R1A2B4010826).

References

1. Korea National Statistical Office: Annual report on the statistical for elderly (2011)
2. Black, D.S., O'Reilly, G.A., Olmstead, R., Breen, E.C., Irwin, M.R.: Mindfulness meditation and improvement in sleep quality and daytime impairment among older adults with sleep disturbances: a randomized clinical trial. JAMA Intern. Med. **175**(4), 494–501 (2015)
3. Jeong, H.S., Song, Y.M.: Contributing factors to the increases in health insurance expenditures for the aged and their forecasts. Korean J. Health Econ. Policy **19**(2), 21–38 (2013)
4. Korea Centers for Disease Control and Prevention: Korea National health & nutrition examination survey (2007–2014)
5. Bae, Y.H., Shin, J.S., Lee, J., Kim, M.R., Park, K.B., Cho, J.H., Ha, I.H.: Association between Hypertension and the prevalence of low back pain and osteoarthritis in Koreans: a cross-sectional study. PloS one, **10**(9) (2015)
6. Song, S., Paik, H.Y., Song, W.O., Song, Y.: Metabolic syndrome risk factors are associated with white rice intake in Korean adolescent girls and boys. Br. J. Nutr. **113**(03), 479–487 (2015)
7. Ha, I.H., Lee, J., Kim, M.R., Kim, H., Shin, J.S.: The association between the history of cardiovascular diseases and chronic low back pain in South Koreans: a cross-sectional study. PloS one **9**(4) (2014)
8. Piao, Y., Piao, M., Park, K., Ryu, K.H.: An ensemble correlation-based gene selection algorithm for cancer classification with gene expression data. Bioinformatics **28**(24), 3306–3315 (2012)
9. Piao, Y., Piao, M., Ryu, K.H.: Multiclass cancer classification using a feature subset-based ensemble from microRNA expression profiles. Comput. Biol. Med. **80**, 39–44 (2017)
10. Lee, D.G., Ryu, K.S., Bashir, M., Bae, J.W., Ryu, K.H.: Discovering medical knowledge using association rule mining in young adults with acute myocardial infarction. J. Med. Syst. **37**(2), 9896 (2013)

11. Bashir, M.E.A., Shon, H.S., Lee, D.G., Kim, H., Ryu, K.H.: Real-time automated cardiac health monitoring by combination of active learning and adaptive feature selection. TIIS **7**(1), 99–118 (2013)

12. Kim, H., Ishag, M.I.M., Piao, M., Kwon, T., Ryu, K.H.: A data mining approach for cardiovascular disease diagnosis using heart rate variability and images of carotid arteries. Symmetry **8**(6), 4 (2016)

13. Mayer, C., Bachler, M., Holzinger, A., Stein, P.K., Wassertheurer, S.: The effect of threshold values and weighting factors on the association between entropy measures and mortality after myocardial infarction in the Cardiac Arrhythmia Suppression Trial (CAST). Entropy **18**(4), 129, 121–115 (2016)

14. Kaur, H., Wasan, S.K.: Empirical study on applications of data mining techniques in healthcare. J. Comput. Sci. **2**(2), 194–200 (2006)

15. Holzinger, A.: Interactive Machine Learning for Health Informatics: When do we need the human-in-the-loop? Brain Inform. **3**(2), 119–131 (2016)

16. Hund, M., Boehm, D., Sturm, W., Sedlmair, M., Schreck, T., Ullrich, T., Keim, D.A., Majnaric, L., Holzinger, A.: Visual analytics for concept exploration in subspaces of patient groups: Making sense of complex datasets with the Doctor-in-the-loop. Brain Inform. **3**(4), 233–247 (2016)

17. Park, H.W., Batbaatar, E., Li, D., Ryu, K.H.: Risk factors rule mining in hypertension: Korean National Health and Nutrient Examinations Survey 2007–2014. In: Computational Intelligence in Bioinformatics and Computational Biology (CIBCB), pp. 1–4 (2016)

18. Yang, Y., Liao, Y., Meng, G., Lee, J.: A hybrid feature selection scheme for unsupervised learning and its application in bearing fault diagnosis. Expert Syst. Appl. **38**(9), 11311–11320 (2011)

19. Hsu, H.H., Hsieh, C.W., Lu, M.D.: Hybrid feature selection by combining filters and wrappers. Expert Syst. Appl. **38**(7), 8144–8150 (2011)

20. Xie, J., Wang, C.: Using support vector machines with a novel hybrid feature selection method for diagnosis of erythemato-squamous diseases. Expert Syst. Appl. **38**, 5809–5815 (2011)

21. Vapnik, V.: The nature of statistical learning (2013)

22. Han, J., Fu, Y.: Attribute-oriented induction in data mining. Advances in knowledge discovery and data mining, pp. 399–421 (1996)

Statistical Analysis of Perinatal Risk Factors for Emergency Caesarean Section

Ibrahim Abou Khashabh[1]([⊠]), Václav Chudáček[2], and Michal Huptych[2]

[1] Department of Cybernetics, Faculty of Electrical Engineering,
Czech Technical University in Prague, Prague, Czech Republic
aboukibr@fel.cvut.cz
[2] Czech Institute of Informatics, Robotics, and Cybernetics,
Czech Technical University in Prague, Prague, Czech Republic

Abstract. Objective: To explore potential risk factors for the emergency caesarean section in term, singleton pregnancies. Methods: A retrospective population based case-control study in term deliveries from the University Hospital in Brno, Czech Republic collected between 2014 and 2016. Cases were deliveries by emergency caesarean section; controls were all others modes of delivery. We excluded elective caesarean from the populations.

Results: In the database of 13769 deliveries, we identified 2178 cases. Univariate and multivariable analysis of clinical features were performed. The following risk factors were associated with emergency caesarean section: Breech presentation OR 21.6 (14.6–30.5), obstructed labor OR 42.7 (28.5–63.9), scar on uterus OR 18 (14.3–22.5), fetal distress OR 6.1 (5.3–7.2) and primipara OR 3.7 (3.1–4.4).

Conclusion: Univariate and multivariable analysis of the data from 13769 deliveries were performed, and significant risk factors were identified increasing the chance of undergoing caesarean section.

1 Introduction

Caesarean section (CS) is the most common major surgical procedure in many parts of the world [1–3]. The determinants of CS are very complex and include not only clinical indications, but also economic and organizational factors, the physicians' attitudes toward birth management, and the social and cultural attitudes of women. Most clinical indications are not absolute, and many are very subjective and culture-bound, so there is significant variability among hospitals and countries with respect to CS rates for particular medical indications [4].

There is a large pool of works investigating the relation of different risk factors related to delivery with CS. Patel et al. [13] published a study on a population of around 12 000 deliveries describing the relation between various risk factors and CS; among those factors were a breech presentation and previous caesarean section. Gareen et al. [8] investigated the relation between maternal age and rate of CS for nulliparous and parous women. Verhoeven et al. [21] investigated factors leading to CS following labor induction in multiparous women, the risk

© Springer International Publishing AG 2017
M. Bursa et al. (Eds.): ITBAM 2017, LNCS 10443, pp. 20–29, 2017.
DOI: 10.1007/978-3-319-64265-9_3

of CS increase with previous preterm delivery, short maternal height and limited dilatation at the start of induction.

A study by Gulmezoglu [5] contributed to the comparison of a labor induction policy versus a policy of expectant management with an equal distribution of cases and controls, mothers who were managed with an induction policy had an 11% decrease in the caesarean delivery rate. Anim-Somuah [6] made a comparison between epidural analgesia in labor and non-epidural analgesia; CS increased 43% due to fetal distress when mothers managed with epidural analgesia during labor. Wendland [7] documented the evaluation of gestational diabetes mellitus and CS; the risk increased 37% when the women have a gestational diabetes mellitus. Penn [9] defined many major indications for CS as repeat caesarean section, dystocia, breech presentation and fetal distress. A population-based screening study by Weiss [10] defined obesity as an independent risk factor for the adverse obstetric outcome and the CS rate was 33.8% for obese, where obesity was defined as body mass index (BMI) greater than 30 and less than 35.

A systematic review study by Poobalan [23] shows the obesity as an independent risk factor for both elective and emergency caesarean, with the risk increased 50% in overweight women (BMI 25–30) and more than double in obese women (BMI 30–35) compared to controls. A study of 287213 pregnancies in London by Sebire [18] shows that maternal obesity increases the risk of delivery by emergency caesarian section, gestational diabetes mellitus, pre-eclampsia, induction of labor.

Buist [19] shows the CS rate was significantly increased with induced labor (21.5% versus 14.9%, p<0.001), the increased rate occurring in nulliparous women. A case-control study in sweden by Cnattingius [15] shows that the risk of CS was increased among nulliparous, previous caesarean delivery [16]. Induction of labor is thought to be associated with an increase in the risk of CS both for nulliparous and multiparous women.

The present contribution aims to investigate confounding risk factors and their effects independently to emergency CS, to understand the multiple factors that can influence the outcome (CS) and that could be used to enhance delivery evaluation as an addition to methods such as automated fetal heart rate analysis [17].

2 Materials and Methods

Our study is a data-driven retrospective population based case-control study using the clinical database of University Hospital in Brno, Czech Republic.

2.1 Dataset

The data were collected between January 2014 and September 2016 using our data collection system, DeliveryBook running at the delivery ward of the University Hospital in Brno since mid-2013. The application allows collection of clinical information in a structured way. Fields are type/value - locked and the majority

Table 1. Features description assessed by emergency CS.

	N	CS – Cases	CS – Controls
		Mean (SD)	Mean (SD)
pH	13336	7.28 (0.08)	7.29 (0.09)
Apgar score 5 min	13769	9.36 (1.16)	9.59 (0.88)
Weight	13756	3411 (498)	3404 (432)
Height	13691	49.7 (2.43)	49.9 (2.05)
Age of mother	13630	33 (5.02)	32 (4.83)
Ist stage	13354	333 (140)	258 (104)
IInd stage	11953	59 (33)	14 (14)
Gravidity	13768	1.82 (1.25)	2.07 (1.31)
Parity	13758	1.45 (0.74)	1.73 (0.90)

of information is input as a selection from predefined lists. A dedicated midwife is responsible for second reading and error corrections on a weekly basis. Clinically relevant parameters about mother, delivery and newborn are collected. The collection and use of the data for scientific reasons was approved by the ethics committee at the University Hospital in Brno.

2.2 Population and Inclusion Criteria

The database consists of 17113 recordings; following conditions had been used: outcome variable is available (CS); gestational age ≥ 37 weeks; singleton pregnancy; no a priory known congenital diseases. Additionally, all deliveries by planned caesarean section were excluded. In total 13769 deliveries were used for CS related analysis. Each available variable is divided into two groups based on indication of CS or all others modes of deliveries.

2.3 Features

A total of 134 features were analysed – with all of them falling into one of the following categories three categories. First, features detected prior to going into labor (e.g. parity, induction of labor, diagnosis related to mother and pregnancy) and their descriptive statistics is shown in Table 1. Second, features occurring within the delivery period (e.g. interventions, length of delivery stages, medications, diagnosis related to delivery). And finally features known after delivery outcome measures (e.g. pH, Apgar score, admittance to the neonatal intensive care unit or seizures).

2.4 Statistical Evaluation

The relation between each feature and the emergency CS outcome was evaluated independently using univariate and multivariable analysis to obtain odds ratios,

Table 2. Overview of selected clinical characteristics with respect to the outcome CS.

	(%)	CS – Cases #(%)	CS – Controls #(%)
Induced delivery	23.4	698 (32.1)	2282 (19.7)
Company at labor	14.3	1471 (67.5)	8827 (76.2)
Sex (male)	17.3	1225 (56.2)	5848 (50.5)
Mother Age > 40	19.7	113 (5.25)	462 (4.03)
Oxytocin	30.2	617 (33.1)	1425 (14.5)
Epidural analg.	25.5	818 (43.9)	2394 (24.4)
Weight < 2500	28.7	89 (4.09)	221 (1.91)
Weight > 4500	23.01	26 (1.19)	87 (0.75)
Primigravida	20.5	1147 (52.7)	4449 (38.4)
Primipara	21.1	1424 (65.5)	5338 (46.1)
Mother Age > 30	22.4	923 (42.9)	3201 (27.9)
Mother Age > 40	36	41 (1.91)	73 (0.64)
Epidural	56.4	1090 (50.1)	844 (7.29)

Table 3. Overview of ICD coded risk factors with respect to emergency CS.

	(%)	CS – Cases #(%)	CS – Controls #(%)
O321: Breech presentation	70.9	178 (8.17)	73 (0.63)
O649: Obstructed labor	79.9	246 (11.3)	62 (0.53)
O160: Hypertension	21.7	28 (1.29)	101 (0.87)
O130: Gest. hypertension	22.6	112 (5.14)	383 (3.30)
O149: Pre-eclampsia	39	64 (2.94)	100 (0.86)
O240: Pre-existing 1 DM	35.7	15 (0.69)	27 (0.23)
O244: Gest. diabetes	18.5	153 (7.02)	675 (5.82)
O342: Scar on uterus	48.9	459 (21.1)	480 (4.14)
O365: IUGR	34	106 (4.87)	206 (1.78)
O366: Excessive fetal growth	47.3	69 (3.17)	77 (0.66)
O420: PROM	14.6	277 (12.7)	1616 (13.9)
O681: Meconium	22.4	272 (12.5)	941 (8.12)
O688: Fetal distress (CTG)	42.5	722 (33.2)	979 (8.45)
O752: Pyrexia intrapartum	38.9	21 (0.96)	33 (0.28)
E669: Obesity	41.7	40 (1.84)	56 (0.48)
G409: Epilepsy	21.5	14 (0.64)	51 (0.44)

Table 4. Univariate selected clinical characteristics with respect to the outcome CS.

	OR (95 % CI)
Induced delivery	1.92 (1.74–2.13)
Company at labor	0.65 (0.59–0.72)
Sex (male)	1.26 (1.15–1.38)
Mother Age > 40	1.32 (1.07–1.63)
Oxytocin	2.91 (2.61–3.26)
Epidural analg	2.43 (2.19–2.69)
Weight < 2500	2.19 (1.71–2.81)
Weight > 4500	1.60 (1.03–2.48)
Primigravida	1.79 (1.63–1.96)
Primipara	2.22 (2.02–2.44)
Mother Age > 30	1.94 (1.77–2.14)
Mother Age > 40	3.04 (2.06–4.46)
Epidural	12.7 (11.4–14.2)

Table 5. Univariate analysis of ICD coded risk factors for emergency CS.

	OR (95 % CI)
O321: Breech presentation	14 (10.7–18.5)
O649: Obstructed labor	23.7 (17.9–31.4)
O160: Hypertension	1.48 (0.97–2.26)
O130: Gest. hypertension	1.59 (1.28–1.97)
O149: Pre-eclampsia	3.48 (2.53–4.78)
O240: Pre-existing 1 DM	2.97 (1.58–5.59)
O244: Gest. diabetes	1.22 (1.02–1.47)
O342: Scar on uterus	6.18 (5.39–7.09)
O365: IUGR	2.83 (2.23–3.59)
O366: Excessive fetal growth	4.89 (3.52–6.79)
O420: PROM	0.90 (0.78–1.03)
O681: Meconium	1.62 (1.40–1.86)
O688: Fetal distress (CTG)	5.38 (4.81–6.00)
O752: Pyrexia intrapartum	3.41 (1.97–5.90)
E669: Obesity	3.85 (2.56–5.80)
G409: Epilepsy	1.46 (0.81–2.65)

their 95% confidence intervals (CI), in addition to basic descriptive statistics. An odds ratio (OR) is a measure of the association between exposure and an outcome. It answers the question - how much of a difference does the presence of the risk factor have on the chance of CS [14]? Odds ratios are most commonly used in case-control studies. However, they can also be used in cross-sectional and cohort study designs [20]. Additional information on the OR can be found in the statistical notes by Bland [24], with arguments supporting the OR over relative risk (RR) estimation. In our case, the difference between OR and RR is negligible, since the event rate of pathological cases is very low, so $OR \simeq RR$ [14].

The interpretation of odds ratios makes use of Borenstein's formula [25] and Cohen's d [26] for effect size such that OR > 1.44, OR > 2.97, and OR > 4.27 represent small, medium, and large effects, respectively [26]. The univariable analyses were performed in Matlab 2015a, and the multivariable analyses were performed in IBM SPSS Statistics suite 22.

3 Results

A total of 13769 deliveries fulfilled the inclusion criteria, of these 2178 (15.8%) underwent Emergency CS. The most common indications for emergency CS are a breech presentation and obstructed labor. Features describing medical history of the mother, main characteristics of the fetus and features related to delivery are presented in Table 2, and related univariate analysis in Tables 4 and 6 depicts in greater detail relation of cases occurrence on day of the week and month – evaluating thus influence of iatrogenic as well as season's influences. Finally more details on age of the mother and body mass index observed in pre-gestation period and during gestation are shown in Table 7.

Table 3 presents ICD-10 codes possibly relevant to Emergency CS with the univariate analysis in Table 5. The overview tables show the percentage of each feature according to emergency CS with the number of cases and controls, among the features that lead to emergency CS we identified: breech presentation, obstructed labor, scar on the uterus (previous CS), obesity and excessive fetal growth. The multivariable model in Table 8 shows important risk factors for emergency CS.

4 Discussion

Several risk factors related to emergency CS outcome were identified. The common indications for CS: breech presentation, obstructed labor, and epidural anesthesia have the highest risks; An explanation might be that breech deliveries are started as vaginal but end up with emergency CS. Many papers investigate the independent association between features and CS that increase the emergency CS rate: induced delivery, primiparous women, epidural, pre-eclampsia, pre-existing DM, scar on the uterus, IUGR, fetal distress (CTG), pyrexia intrapartum and obesity [2, 18, 23, 27].

Table 6. Relation of Acute CS and days months per cases and controls.

	CS – Cases (%)	CS – Controls #(%)	OR (95 % CI) #(%)
Monday	240 (15.68)	1291 (84.32)	0.98 (0.84–1.14)
Tuesday	277 (17.13)	1340 (82.87)	1.11 (0.96–1.28)
Wednesday	289 (17.38)	1374 (82.62)	1.13 (0.99–1.30)
Thursday	254 (15.55)	1379 (84.45)	0.97 (0.84–1.12)
Friday	238 (15.40)	1307 (84.60)	0.96 (0.82–1.11)
Saturday	205 (15.93)	1082 (84.07)	1.00 (0.85–1.17)
Sunday	180 (14.46)	1065 (85.54)	0.88 (0.74–1.04)
Holidays	38 (13.29)	248 (86.71)	0.80 (0.57–1.14)
January	189 (16.22)	976 (83.78)	1.03 (0.87–1.21)
February	166 (15.00)	941 (85.00)	0.92 (0.78–1.10)
March	145 (15.57)	786 (84.43)	0.97 (0.81–1.17)
April	123 (15.19)	687 (84.81)	0.94 (0.77–1.15)
May	140 (16.47)	710 (83.53)	1.04 (0.86–1.26)
June	154 (17.95)	704 (82.05)	1.17 (0.97–1.40)
July	136 (14.18)	823 (85.82)	0.86 (0.71–1.04)
August	135 (15.66)	727 (84.34)	0.98 (0.81–1.19)
September	144 (16.51)	728 (83.49)	1.05 (0.87–1.26)
October	125 (15.34)	690 (84.66)	0.95 (0.78–1.16)
November	133 (17.03)	648 (82.97)	1.09 (0.90–1.32)
December	131 (16.44)	666 (83.56)	1.04 (0.86–1.27)

This study highlights two points. Maternal age of more than 40 years has a small effect OR 1.32 (1.07–1.63) to emergency CS, but risk for primiparous women over 40 years old has a medium effect OR 3.04 (2.06–4.46). Therefore primipara with increased maternal age is an important risk factor for emergency CS.

The diagnosis O366: Excessive fetal growth that is diagnosed before delivery shows a large effect to emergency CS with OR 4.89 (3.52–6.79), while the increasing of newborn weight over 4500 g shows small effect with OR 1.60 (1.03–2.48), which means that there might be inconsistencies in the evaluation of the excessive fetal growth. Both excessive fetal growth and newborn weight did not appear as significant factors in multivariable regression.

Although the size of the database is promising, the main concern is the size of the subsets representing individual features. Especially for diagnostic features where ICD-10 codes are used, many subsets contain only tens of cases, which is insufficient to allow conclusions to be drawn with high confidence.

Table 7. In detail stratification of age, and BMI pre-gestation and during gestation for cases and controls of the Acute CS.

	CS – Cases (%)	CS – Controls #(%)	OR (95 % CI) #(%)
Maternal age (20–25)	122 (15.10)	686 (84.90)	0.94 (0.77–1.14)
Maternal age (25–30)	474 (17.21)	2281 (82.79)	1.14 (1.01–1.28)
Maternal age (30–35)	782 (15.15)	4381 (84.85)	0.90 (0.81–0.99)
Maternal age (35-40)	517 (16.01)	2712 (83.99)	1.01 (0.90–1.13)
Maternal age < 20	15 (15.31)	83 (84.69)	0.96 (0.55–1.66)
Maternal age > 40	98 (20.04)	391 (79.96)	1.35 (1.07–1.69)
BMI pre-gest. (20–25)	848 (14.81)	4876 (85.19)	0.84 (0.76–0.93)
BMI pre-gest. (25–30)	365 (18.71)	1586 (81.29)	1.28 (1.12–1.45)
BMI pre-gest. (30–35)	152 (22.93)	511 (77.07)	1.63 (1.35–1.97)
BMI pre-gest. (35–40)	47 (26.40)	131 (73.60)	1.92 (1.37–2.69)
BMI pre-gest. < 20	287 (13.11)	1903 (86.89)	0.76 (0.66–0.87)
BMI pre-gest. > 40	17 (19.54)	70 (80.46)	1.29 (0.76–2.19)
BMI gest. (20–25)	290 (11.14)	2313 (88.86)	0.59 (0.52–0.68)
BMI gest. (25–30)	732 (14.52)	4310 (85.48)	0.82 (0.74–0.91)
BMI gest. (30–35)	472 (21.02)	1774 (78.98)	1.56 (1.39–1.76)
BMI gest. (35–40)	149 (23.84)	476 (76.16)	1.72 (1.42–2.08)
BMI gest. < 20	3 (6.82)	41 (93.18)	0.39 (0.12–1.25)
BMI gest. > 40	71 (30.47)	162 (69.53)	2.37 (1.79–3.15)

Table 8. Multivariable model of risk factors for emergency CS.

	OR (95 % CI)
O321: Breech presentation	21.1 (14.6–30.5)
O649: Obstructed labor	42.7 (28.5–63.9)
O342: Scar on uterus	18 (14.3–22.5)
O688: Fetal distress (CTG)	6.1 (5.3–7.2)
Primipara	3.7 (3.1–4.4)

5   Conclusion

A univariate analysis and multivariable regression of risk factors for emergency CS was undertaken and identified several features associated with emergency CS outcome. The univariate analysis showed the following effects: Small effect - Induced delivery, advanced maternal age, and excessive newborn weight. Medium effect - Oxytocin, pre-eclampsia, primiparous with advanced maternal age and obesity. Large effect - Fetal distress (CTG), scar on the uterus and excessive fetal growth.

The final multivariable model showed the following risks factors: Breech presentation, obstructed labor, Scar on the uterus, fetal distress and primipara. The results imply that there is a need for accurate screening of women during maternal care and, decision to perform CS should be based on clear, compelling and well-supported justifications.

Acknowledgment. This work was supported by SGS grant SGS17/216/OHK4/3T/37 of the CTU in Prague. Additionally we would like to thank our clinical colleagues, namely Dr. Petr Janků and Dr. Lukáš Hruban for their meticulous collection and interpretation of the data.

References

1. National Collaborating Centre for Women's and Children's Health. Caesarean section Guidline. Royal College of Obstetricians and Gynaecologists. 27 Sussex Place: RCOG Press (2004)
2. Shamshad, B.: Factors leading to increased cesarean section rate. Gomal. J. Med. Sci. **6**(1), 1–5 (2008)
3. Gibbons, L., et al.: Inequities in the use of cesarean section deliveries in the world. Am. J. Obstet. Gynecol. **6**(1), 1–5 (2008)
4. Arrieta, A.: Health reform and cesarean sections in the private sector: the experience of Peru. Health Policy **99**(2), 124–130 (2011)
5. Gulmezoglu, A. Metin et al.: Induction of labour for improving birth outcomes for women at or beyond term. In: The Cochrane library (2012)
6. Anim-Somuah, M., Smyth, R.M., Jones, L.: Epidural versus non-epidural or no analgesia in labour. Cochrane Database Syst. Rev. **12**, CD000331 (2011)
7. Wendland, E.M., et al.: Gestational diabetes and pregnancy outcomes-a systematic review of the World Health Organization (WHO) and the International Association of Diabetes in Pregnancy Study Groups (IADPSG) diagnostic criteria. BMC Pregnancy Childbirth **12**(1), 23 (2012)
8. Gareen, I.F., Morgenstern, H., Greenland, S., Gifford, D.S.: Explaining the association of maternal age with cesarean delivery for nulliparous and parous women. J. Clin. Epidemiol. **56**(11), 1100–1110 (2003)
9. Penn, Z., Ghaem-Maghami, S.: Indications for caesarean section. Best Pract. Res. Clin. Obstet. Gynaecol. **15**(1), 1–15 (2001)
10. Weiss, J.L., et al.: Obesity, obstetric complications and cesarean delivery rate-a population-based screening study. Am. J. Obstet. Gynecol. **190**(4), 1091–1097 (2004)
11. Lepercq, J., et al.: Factors associated with cesarean delivery in nulliparous women with type 1 diabetes. Obstet. Gynecol. **115**(5), 1014–1020 (2010)
12. Abebe, F.E., et al.: Factors leading to cesarean section delivery at Felegehiwot referral hospital, Northwest Ethiopia: a retrospective record review. Reprod. Health **13**(1), 6 (2016)
13. Roshni, R.P., Tim, J.P., Deirdre, J.M., ALSPAC Study Team, et al.: Prenatal risk factors for Caesarean section. Analyses of the ALSPAC cohort of 12 944 women in England. Int. J. Epidemiol. **34**(2), 353–367 (2005)

14. Schmidt, C.O., Kohlmann, T.: When to use the odds ratio or the relative risk? Int. J. Public Health **53**(3), 165–167 (2008)
15. Cnattingius, R., Höglund, B., Kieler, H.: Emergency cesarean delivery in induction of labor: an evaluation of risk factors. Acta Obstet. Gynecol. Scand. **84**(5), 456–462 (2005)
16. Yeast, J.D., Jones, A., Poskin, M.: Induction of labor and the relationship to Cesarean delivery: a review of 7001 consecutive inductions. Am. J. Obstet. Gynecol. **180**(3), 628–633 (1999)
17. Doret, M., Spilka, J., Chudacek, V., et al.: Fractal analysis and hurst parameter for intrapartum fetal heart rate variability analysis: a versatile alternative to frequency bands and LF/HF ratio. PLOS ONE **10**(8), e0136661 (2015)
18. Sebire, N.J., et al.: Maternal obesity and pregnancy outcome: a study of 287 213 pregnancies in London. Int. J. Obesity **25**(8), 1175 (2001)
19. Buist, R.: Induction of labour: indications and obstetric outcomes in a tertiary referral hospital. N. Z. Med. J. **112**(1091), 251–253 (1999)
20. Szumilas, M.: Explaining odds ratios. J. Can. Acad. Child Adolesc. Psychiatry **19**, 227 (2010)
21. Verhoeven, C.J., van Uytrecht, C.T., Porath, M.M., Mol, B.W.J.: Risk factors for cesarean delivery following labor induction in multiparous women. J. Pregnancy (2013)
22. Szal, S.E., Croughan-Minihane, M.S., Kilpatrick, S.J.: Effect of magnesium prophylaxis and preeclampsia on the duration of labor. Am. J. Obstet. Gynecol. **180**(6), 1475–1479 (1999)
23. Poobalan, A.S., et al.: Obesity as an independent risk factor for elective and emergency Caesarean delivery in nulliparous women-systematic review and meta-analysis of cohort studies. Obes. Rev. **10**(1), 28–35 (2009)
24. Bland, J.M., Altman, D.G.: The odds ratio. BMJ **320**(7247), 1468 (2000)
25. Borenstein, M., Hedges, L.V., Higgins, J.P.T., Rothstein, H.R.: Introduction to Meta-Analysis. Wiley, Chichester
26. Cohen, J., McIntire, D.D., Leveno, K.J.: Statistical Power Analysis for the Behavioral Sciences, 2nd ed., pp. 281, 284, 285. Erlbaum., Hillsdale
27. Hannah, M.E., et al.: Planned caesarean section versus planned vaginal birth for breech presentation at term: a randomised multicentre trial. Lancet **356**(9239), 1375–1383 (2000)

Modelling of Cancer Patient Records: A Structured Approach to Data Mining and Visual Analytics

Jing Lu[1(✉)], Alan Hales[1,2], and David Rew[2]

[1] University of Winchester, Winchester SO22 5HT, UK
Jing.Lu@winchester.ac.uk
[2] University Hospital Southampton, Southampton SO16 6YD, UK
aahales@btinternet.com, D.Rew@soton.ac.uk

Abstract. This research presents a methodology for health data analytics through a case study for modelling cancer patient records. Timeline-structured clinical data systems represent a new approach to the understanding of the relationship between clinical activity, disease pathologies and health outcomes. The novel Southampton Breast Cancer Data System contains episode and timeline-structured records on >17,000 patients who have been treated in University Hospital Southampton and affiliated hospitals since the late 1970s. The system is under continuous development and validation. Modern data mining software and visual analytics tools permit new insights into temporally-structured clinical data. The challenges and outcomes of the application of such software-based systems to this complex data environment are reported here. The core data was anonymised and put through a series of pre-processing exercises to identify and exclude anomalous and erroneous data, before restructuring within a remote data warehouse. A range of approaches was tested on the resulting dataset including multi-dimensional modelling, sequential patterns mining and classification. Visual analytics software has enabled the comparison of survival times and surgical treatments. The systems tested proved to be powerful in identifying episode sequencing patterns which were consistent with real-world clinical outcomes. It is concluded that, subject to further refinement and selection, modern data mining techniques can be applied to large and heterogeneous clinical datasets to inform decision making.

Keywords: Clinical data environment · Electronic patient records · Health information systems · Data mining · Visual analytics · Decision support

1 Introduction

The healthcare industry has many established systems being used for electronic patient records, hospital administration, resource management and to circulate clinical results. There is a growing need to be able to share large amounts of health data, perform complex analysis and visualise lifeline tracks of patients. One of the latest approaches is through the implementation of a digital strategy at various levels.

© Springer International Publishing AG 2017
M. Bursa et al. (Eds.): ITBAM 2017, LNCS 10443, pp. 30–51, 2017.
DOI: 10.1007/978-3-319-64265-9_4

1.1   Background

After years of digitising patient records, the UK National Health Service (NHS) is acquiring a considerable repository of clinical information – hundreds of millions of test results and documents for tens of millions of patients. NHS England describes the role of health informatics as being fundamental to the transformational changes needed. As a result the NHS is investing in a number of initiatives to integrate data and provide insight. However, despite this, the NHS "has only just begun to exploit the potential of using data and technology at a national or local level" [17].

While the NHS collects huge amounts of data on patient care, it is nevertheless very difficult to establish ground truths in the relationships between multidisciplinary clinical inputs and therapeutic outcomes in a wide variety of chronic diseases of childhood and adulthood. This is not only because of the sheer complexity of human lives and populations, but also because of the number of variables which affect real-world health events. All human lives and health events play out over time, from conception to death, such that the passage of time is a central element in each and every disease process. Despite this truism, it is not easy to find commercial health informatics systems in which temporally-structured data presentation and analysis is a central element.

The electronic patient record (EPR) system is a digital compilation of every patient's healthcare information, unique identifiers and demographic data, and contains a range of documents from notes to test results [22]. One benefit to having EPRs is that paper records no longer have to be maintained, which supports the government's 2018 vision of a paperless NHS.

Emerging big data technologies mean it is now possible to share data from systems that previously could not communicate; this could potentially allow different parts of the health service to work together [18]. All of this stored data can be used for aiding decisions or to learn something new [15]. One approach which could provide doctor's with the ability to derive useful information from massive medical datasets is known as big data analytics.

1.2   Case Study – University Hospital Southampton

In 1996, Professor Ben Shneiderman and colleagues at the Human Computer Interaction Laboratory in Bethesda at the University of Maryland developed a conceptual structure for the visualisation of clinical records in which a complex and heterogeneous series of data could be displayed and easily understood on parallel timelines within a simple two-dimensional graphic. This nevertheless had powerful features and it formed the basis for a universal descriptor for human lives and medical histories as a tool for the overview of a complex dataset, into which individual documents and reports could be readily zoomed, and from which extraneous or unwanted detail could be temporarily filtered out [14].

The Lifelines model has been adopted as a framework for a practical, dynamic and interactive EPR which now sits within the University Hospital Southampton (UHS) clinical data environment (CDE). It provides much faster access times and overviews to the more than one million patient records compared with other available software.

UHS Lifelines has proved to be a valuable testbed for timeline-structured EPRs and it remains in continuous, agile and iterative development.

Figure 1 is a screenshot of the UHS Lifelines timeline-structured EPR graphical interface, taken from a patient record within the Southampton Breast Cancer Data System. This figure illustrates a number of unique features of the system including separate clinical and reporting timelines, on which are displayed icons at the time of their generation in an intuitive manner. Clicking on each icon displays the underlying document or report. The lowermost timeline, labelled "Cancer Events", is the master timeline or "UHS Lifetrak". In this case, a patient developed a right-sided cancer in December 2014 (green circle) and overt metastases in January 2016 (yellow triangle).

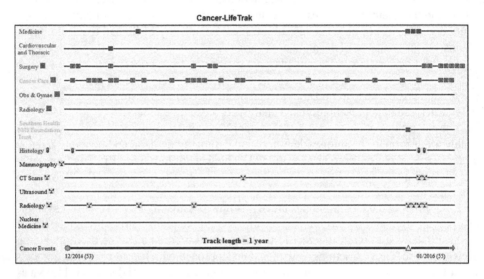

Fig. 1. Screenshot of the UHS Lifelines timeline-structured EPR graphical interface (Color figure online)

The clinical document outputs and diagnostic test result events shown are those which are most frequently relevant and informative when considering a breast cancer case. The entire diagnostic test result history for a patient is frequently so large however that it would create visual noise and severely limit the effectiveness of the lifeline graphic.

The data visualisation and access tools within UHS Lifelines provide a methodology which can be adapted to research into the natural history and progression of a wide range of human diseases, and the opportunity for new approaches to temporally-orientated data mining and data analysis [22]. This paper presents a framework in this context and describes the application of several data mining tools to develop new insights into the data and the structure of the data system.

Since 2012 UHS have designed, built and populated the Southampton Breast Cancer Data System (SBCDS) to help understand the clinical inputs and outcomes for a substantial cohort of breast cancer patients who have been treated in one of the largest specialist centres in the UK since the 1970s. A variety of data sources has been used including a card index of more than 12,000 patients, which has been maintained continuously since 1978, and a range of legacy concurrent datasets with information on breast cancer patients.

It soon became apparent that, by building links to elements of the UHS Lifelines data system into the individual cancer patient records, there would be a direct evidence base for each patient's clinical progression within the EPR system. This evidence base would accrue continuously to the patient record (when alive) at every contact with the hospital, in whatever clinical discipline. By integrating access to the electronic documents into the data system, the cost and time penalty of populating the individual datasets would be reduced significantly when compared with calling forward paper records.

The result has been that from 2012 onwards SBCDS has accrued more than 17,000 unique timeline-structured patient records, which continue to accrue at the rate of around 10 per week. These are undergoing continuous updating and validation, the results of which will be published separately in due course.

It then became apparent that the clinical progression of the disease could be described for each and every patient along a master timeline, which has been called the UHS Cancer-LifeTrak – because it resembles the single track railway line on a UK Ordnance Survey Land Ranger map. Each station on the line would represent a point of transition in the patient's care pathway – from the time of Primary Diagnosis through to Loco-Regional Recurrence to the appearance of overt distant Metastases to final outcome. In practice, the progression of disease is very variable and complex, and its representation is more challenging because patients may present with left, right or bilateral tumours in series or parallel. As Southampton is a maritime city, the right-sided tumours are represented by green icons (starboard buoys) and left-sided tumours by red icons (port buoys).

The fact of these episode markers now allows the duration of episodes to be measured between the various phase transitions of the disease. These time intervals can be related back to the original pathological descriptors of the tumours and to the multidisciplinary treatment inputs, which variously include surgery, systemic chemotherapy, radiotherapy and anti-oestrogenic therapy. The time points are measured in months and years (mm/yyyy), as measuring by days would afford spurious accuracy to the data when diagnoses and treatments occur over days, weeks and months.

The temporal structure to the data system presented an opportunity to explore a range of different tools for data mining and analytics. This motivated a collaborative research project between UHS and Southampton Solent University, which began in 2014 with the following objectives: enhancement of the SBCDS user interface; expansion of its data mining capability; and exploitation of large-scale patient databases. Anonymised breast cancer data from UHS was pre-processed and several data mining techniques were implemented to discover frequent patterns from disease event sequence profiles [12].

The remainder of this paper proceeds as follows: a process-driven framework for health data analytics is proposed in Sect. 2 which comprises a data layer, functional layer and user layer. The data warehousing, data mining and visualisation components of the methodology are discussed further. In Sect. 3, following a description of the data sources, the emphasis is on the pre-processing for data mining and multi-dimensional modelling. A series of results is given in Sect. 4 which covers the visual analytics and data mining techniques, highlighting sequential patterns graph and decision tree classification. The discussion then proceeds to include an evaluation by domain experts before the concluding remarks.

2 Methodology

Informed by the collaborative work using a complex anonymised real-life dataset, a process-driven framework for health data analytics is described below.

2.1 Process-Driven Framework

Healthcare generates a vast amount of complex data with which to support decision making through information processing and knowledge extraction. The growing amount of data challenges traditional methods of data analysis and this has led to the increasing use of emerging technologies – a conceptual architecture is shown (see Fig. 2) which integrates the pre-processing, data warehousing, data mining and visualisation aspects. In the healthcare context, the data may include clinical records, medical and health research records, administrative and financial data – for example Hospital Episode Statistics, Summary Care Records, the Doctor's Work List and the Patient Administration System.

The process-driven framework comprises a data layer, a functional layer and a user layer. Pre-processing is used as part of the data layer in order to clean raw data and prepare the final dataset for use in later stages. Data cleansing and preparation stages include basic operations such as removing or reducing noise, handling outliers or missing values, and collecting the necessary information to model. Extraction, Transformation and Loading (ETL) is well known throughout the database industry: extracting data from various sources then transforming it through certain integration processes before finally loading the integrated data into a data warehouse. The data from the warehouse is held in a structured form and available for data mining, visualisation or analytics.

The functional layer includes (1) Data Warehousing – integrating data from multiple healthcare systems to provide population-based views of health information; (2) Visual Analytics – applying data visualisation techniques to healthcare data, transforming clinical information into insight through interactive visual interfaces; (3) Data Mining – bringing a set of tools and techniques that can be applied to large-scale patient data to discover underlying patterns, providing healthcare professionals an additional source of knowledge for making decisions; and (4) Timeline Visualisation – comprising a patient lifeline system with application to chronic diseases, enabling tracking of clinical/patient events over time.

Finally the user layer shows the possible results which can be derived, including for example graphical charts for survival analysis, representing output from visual analytics; as well as sequential patterns graphs and decision trees from data mining. These will be illustrated in the Results section of this paper.

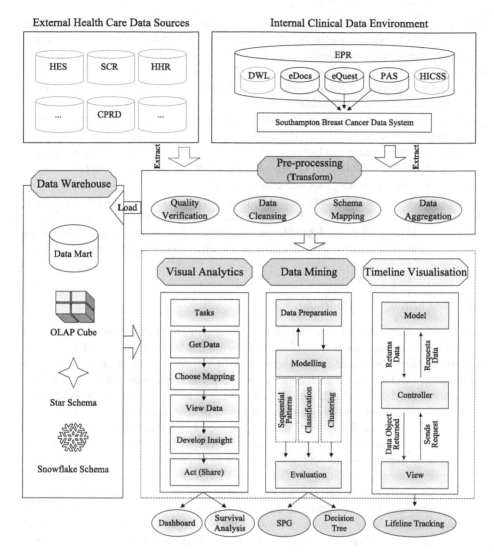

Fig. 2. Process-driven framework for health data analytics

2.2 Data Warehousing

Healthcare datasets come from various sources while health information systems are generally optimised for high speed continuous updating of individual patient data and patient queries in small transactions. Using data warehousing can integrate data from multiple operational systems to provide population-based views of health information.

A data warehouse can be defined as "a copy of transaction data specifically structured for query and analysis" [9]. In order to facilitate decision-makers, complex information systems are assigned with the task of integrating heterogeneous data deriving from operational activities. Case studies include one from Stolba and Tjoa [23], who used a clinical evidence-based process model for the generation of treatment rules. Another example, the Data Warehouse for Translational Research (DW4TR), has been developed to support breast cancer translational research in the USA and this has been extended to support a gynaecological disease programme [7].

A data warehouse is often a collection of data marts: a sub-set of a data warehouse containing data from just one subject area. There are several ways a data warehouse or data mart can be structured, for example multi-dimensional, star or snowflake. However the underlying concept used by all the models is that of a *dimension*, representing the different ways information can be summarised such as by geography, time intervals, age groups and patients. Common to the star and snowflake models is the *fact* table, which contains data (factual history) such as cost or quantity.

On-Line Analytical Processing (OLAP) is an approach to answering multi-dimensional analytical queries. An OLAP cube is a term that typically refers to multi-dimensional arrays of data. OLAP tools enable users to analyse data interactively from multiple perspectives and consist of analytical operations such as roll-up, drill-down, and slicing and dicing.

2.3   Data Mining and Modelling

Data mining is the essential part of *knowledge discovery in databases* – the overall process of converting raw data into useful information and derived knowledge – one definition being "the science of extracting useful information from large datasets or databases" [5]. Data mining techniques could be particularly useful in healthcare and personalised medicine through the following areas of activity: drug development and research, forecasting treatment costs and demand of resources, anticipating patients' future behaviour given their history and the usage of data mining for diagnosis [6].

While data preparation will be discussed further in Sect. 3.2, three data mining methods have been considered: Sequential Patterns Mining aims to find sub-sequences that appear *frequently* in a sequence database; Classification maps each data element to one of a set of pre-determined classes based on the *differences* among data elements; Clustering divides data elements into different groups based on the *similarity* between elements within a single group. Once a model is built from a data analysis perspective, it is important to evaluate the results and review the steps executed to construct the model.

Regarding classification applications in breast cancer studies, Jerez-Aragones et al. [8] presented a decision support tool for the prognosis of breast cancer relapse. It combined an algorithm for selecting the most relevant factors for prognosis of breast cancer with a system composed of different neural network topologies. The identification of breast cancer patients for whom chemotherapy could prolong survival has been treated as a data mining problem as well [10]. Martin et al. [16] examined factors related to the type of surgical treatment for breast cancer using a classification approach and Razavi et al. [19] discuss a decision tree model to predict recurrence of breast cancer.

Clustering techniques have also been applied in breast cancer diagnosis with either benign or malignant tumours [2]. The comparison from their results showed that the k-means algorithm gave an optimum outcome due to better accuracy. In addition, sequential patterns mining has been explored to show the applicability of an alternative data mining technique, e.g. its application to a General Practice database to find rules involving patients' age, gender and medical history [21].

2.4 Visualisation

Visual Analytics

Visual analytics is an integrated approach that combines data analysis with data visualisation and human interaction. There are four separate stages in the process – data, visualisation, knowledge and models. Data mining methods are often used to generate models of the original data. Visual analytics normally commences with a pre-determined task – then goes through an iterative process to get the required data, choose appropriate visual structure (e.g. chart/graph), view the data, formulate insight and then act. This process involves users moving around between different steps as new data insights (and new questions) are revealed.

Tableau Software (https://www.tableau.com) supports this iterative process and provides a collection of interactive data visualisation products focused on business intelligence. After connecting to the data warehouse and adding the tables needed for the analysis, Tableau identifies the fact and dimension tables then sets up every dimension for immediate analysis. Sample results will be demonstrated in Sect. 4.1.

Timeline Visualisation

It has been recognised that the timeline-based data visualisation model can be used as a generic tool with application in the study of all chronic diseases of childhood and adulthood, and as a template for other forms of health informatics research. The concept of the Lifelines EPR can thus been extended to the development of an integrated data system within the UHS-CDE using breast cancer as an exemplar [22].

The model–view–controller (MVC) architectural pattern and a timeline visualisation library can be applied to implement user interfaces. As shown in Fig. 2, within the Timeline Visualisation box, MVC divides a software application into three interconnected parts and each of them are built to handle specific development aspects [20]. Model represents the data structure and typically contains functions to retrieve, insert and update the information in the database. View is the information that is being presented to a user and it is normally via a web page. Controller serves as an intermediary between model and view – it handles user requests and retrieves data.

MVC is one of the most frequently used industry-standard web development frameworks to create scalable and extensible projects. Considering two of the models in this case study – the patient model and the events model – based on patient ID, the controller will open up the patient model and subsequently retrieve all the relevant events. The events model will prepare the data for the template and return it back to the patient model, which will then return that prepared data back to the controller. The controller will then pass that data to the view to visualise the timeline.

3   Clinical Data

The data sources from the Southampton case study are described in this section along with the pre-processing and multi-dimensional modelling techniques.

3.1   Data Sources and Understanding

Extracting data from the clinical environment requires knowledge of the database model and data dictionary as well as domain expertise. In this study, UHS data has been extracted in November 2015 from SBCDS which contains information for 17,000+ breast cancer patients, with a total of 23,200+ records (instances) showing their cancer details. The principle to extract data for this research project is to strictly avoid providing sufficient information that an individual might reasonably be identified by a correlation of seemingly anonymous individual items. Based on this principle, four tables have been exported that reflect SBCDS structure and were loaded into an Oracle database: Patient_Master, Cancel_Detail, Attribute_Data and Attribute_Decodes.

In addition, two more tables have been added to help demonstrate the event clustering challenge through overloading of the graphical interface: Clinical_Events and Event_Codes. Figure 3 shows the ERD for the six tables extracted from SBCDS – note that Request_Type and Request_Code in the Event_Codes table correspond to Specialty and Event_Context in Clinical_Events. This model does not represent the original system which also includes data entry, system management and data analysis.

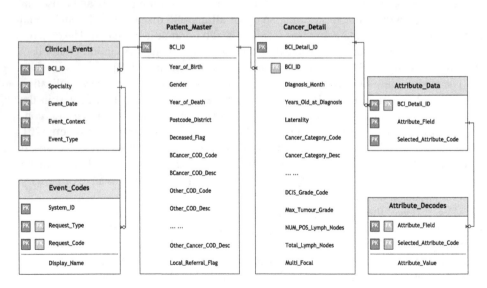

Fig. 3. Entity Relationship Diagram (ERD) for tables extracted from SBCDS

It can be helpful to show the other types of clinical data that breast cancer patients often have, and how that data relates temporally to the cancer data. This data would be deliberately thin in terms of attributes to avoid any confidentiality issues for patients.

A set of decode values has been used to make sense of the coded values that accompany the patient ID and the event date. The events include a mixture of pathology, radiology and clinical documentation. This data is for a small number of patients only – if the same set of data was extracted for all breast cancer patients, it would come to in excess of a million rows of event data.

3.2 Data Pre-processing

Real-world data is occasionally incomplete (with no values for some variables), noisy (with errors and outliers) and inconsistent (with different categorisation or scale). Data pre-processing aims to prepare the data for data mining, modelling and analytics [4]. Good quality data can be achieved by cleansing which is the process of detecting, correcting or removing corrupt or inaccurate values from a record, table or database. This refers to identifying incomplete, incorrect, inaccurate or irrelevant parts of the data and then replacing, modifying or deleting the dirty data [13].

General Data Issues

Table 1 shows some general sample issues and solutions at the data pre-processing stage. As examples of outliers in the datasets, some patients were born before 1900 but are still considered to be alive and other patients do not have an initial record of DoB. Some deceased patients have been recorded with their year of death before 1920. A solution here is to consider the Year_of_Birth (e.g.) after 1920 as a cut off, thus to remove those patients and their corresponding cancer records.

Table 1. Anomaly properties and detection methods

Table name	Data issues	Action required
Patient_Master	Year_of_Birth is unknown	Exclude the records for analysis which need patient's age (group)
	Year_of_Birth range	Exclude the patients (e.g.) born before 1920
	Deceased_Flag = 'N' but Year_of_Death is certain	Replace Deceased_Flag by 'Y'
Cancer_Detail	Laterality	Can be determined by further checking the value of tumour size on "L" or "R" side
	Missing Cancer_Category	Exclude the records for analysis which need cancer type
Attribute_Data	Attribute_Field = 'Cancer_Surgery_L/R'	Exclude the records for the surgery analysis which have both simple mastectomy and wide local excision on the same date
	Attribute_Value = 'Simple mastectomy' OR 'Wide local excision'	

Incomplete data is an unavoidable problem in dealing with most real-world data sources. Simply ignoring the instances is not the best way to handle the missing data and the method which has been used here is to identify the most similar attributes to the one with a missing value and deduce the case for it. For example the Deceased_Flag and Year_of_Death will together indicate if the patient has died – sometimes it is known that a patient has died without knowing the date of death with certainty – however, the Deceased_Flag should be "Y" if there is a valid value for Year_of_Death.

Laterality identifies the breast side diagnosed with cancer, with values of: L = Left; R = Right; B = Bilateral (i.e. both sides); 9 = Unknown (suggesting that the definitive diagnosis report is not available, as any valid diagnosis would identify the side); Null = No value was selected during data entry. There are 78 records with tumour size values for both sides yet the Laterality is not "B" – these have been corrected. In addition, there are 26 instances where the Laterality is "9"; however, further checking of tumour size values suggests that 25 records should be "L" and one should be "R".

Specific Data Issues

As a different data pre-processing example, the success of conservative primary breast cancer surgery (wide local excision) can be compared with radical breast cancer surgery (simple mastectomy). To achieve this a suitable cohort of patients has to be extracted from the SBCDS dataset, with the first step to define a set of criteria to ensure that only appropriate/comparable records are extracted. The next step is to retrieve the required data using SQL statements. During this process, some previously unknown erroneous data was found, e.g. some data suggested that patients could have multiple surgeries on the same date – either multiple occurrences of the same surgery or one surgery type followed by the other. For example, within the same day, one patient has received the wide local excision surgery on the left side and simple mastectomy on the right side.

This issue highlighted episodic attribute data that was orphaned by the original system. It was caused by initial entry (e.g.) on the left side followed by a change to the laterality field and entry of data into the other (right) side, but without removal of the data from the side originally entered (i.e. left). When the record was saved, the spurious data on the unselected side was stored along with the attribute data for the intended side. Roughly about 0.1% records have more than two cases of simple mastectomy or wide local excision on a specific date – this erroneous data can be updated in the database system by use of suitable SQL queries – for the purpose of data analysis in this study, it was decided to simply eliminate this group of data entries.

Normalisation

Breast cancer data represents a very tough challenge to analyse and present in a way that is not confusing and potentially misleading. The data needs a degree of normalisation (i.e. establishing subsets of data that are logical to compare and contrast) before much sense can be made of it. For example, it can be questionable trying to perform analysis on data for deceased and alive patients together and significant thought must be given to what certain event profiles actually mean. It is straightforward to break down the data into distinct groups for further analysis and interpretation. For example, those patients (1) who are still alive; (2) who have died and cancer has been assigned as the cause of death; and (3) who have died, but due to causes that are not considered to be related to cancer.

A first step in making sense of these groups is to factor in the length of time since initial diagnosis. More complex analysis would involve looking at whether longer survival correlates with certain patterns of treatment. Comparing patients based on age group and treatment packages is definitely of interest and one of the most difficult challenges is how to summarise the data to make it understandable without hiding potentially valuable information. The comparison and distribution of survival periods are analysed in the Results section based on the different groups.

3.3   Multi-dimensional Modelling

The data to be warehoused includes anonymised patient records and diagnosis records from SBCDS, which also has direct searches of further information from the UHS-CDE, e.g. eDocs, eQuest and the Patient Administration System. It has been extracted from the breast cancer data system and transformed into patient master and cancer detail tables, before loading to the university-hosted data warehouse.

Design and implementation of the data structure is the first step in data warehousing and one of the key aspects for healthcare warehouse design is to find the right scope for different levels of analysis. Questions to ask include which tables are needed and how are they connected to each other. A snowflake schema has been designed initially based on SBCDS and is shown in Fig. 4, where the granularity comes from patient events during hospital visits. The schema employs the original structure of the SBCDS data, which contains information about the patient, treatment, time and age etc. These groups

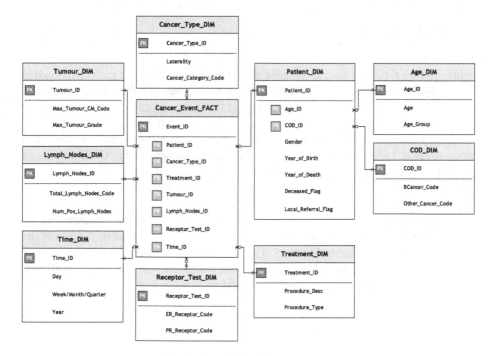

Fig. 4. Snowflake Schema example

are turned into dimensions, as seen in Fig. 4, in order to allow an easy way to analyse the data.

In this schema the available data is divided into measures – fixed facts which can be aggregated – and dimensions, which provide ways to filter the available data. It results in one fact table deriving from cancer events and multiple dimension tables and, when visualised, the fact table is usually in the middle – the schema can be viewed as a snowflake (Fig. 4), hence the name [9].

This data warehouse model has facilitated the exchange of health-related information among the Solent Health Informatics Partnership for research purposes. The data in the warehouse is held in a structured form and available for analysis through OLAP and Tableau. After implementing the design, a working data warehouse has been created. Users are then able to look at the summarised or aggregated data to various levels – through joining the fact table to the selected dimension tables (e.g. patient, treatment, time, age, tumour, cancer_type etc.).

4   Results

This section highlights what can be achieved through visual analytics and data mining on the SBCDS dataset, in particular by using the Tableau and Weka software. These results are indicative only, serving to illustrate the general approach and techniques described. The focus for data mining and modelling will be on sequential patterns mining and classification.

4.1   Visual Analytics

Dashboard
Dashboards are typically used as a means of displaying live data and each dashboard is a collection of individual indicators, designed in such a way that their significance can be grasped quickly. Figure 5 gives an example of the diagnosis dashboard from Tableau and contains five charts each demonstrating a different situation.

The first chart at the top left of Fig. 5 shows the data distribution for both male and female patients. Indeed men can develop breast cancer as well, in the small amount of breast tissue behind their nipples [1]. There are fewer than 70 male patients with about 100 records in the dataset. The study here focuses on female patients, of mean age 61 years old at the diagnosis.

The second chart at the top right of Fig. 5 shows the overall data distribution for deceased and alive patients, where the earliest patient was diagnosed in 1960s and the most recent one was in November 2015. The rest of the dashboard shows other general information about the patients recorded in the data warehouse, i.e. laterality of the breast cancer, percentage of alive/deceased patients and probability that cancer is the cause of death.

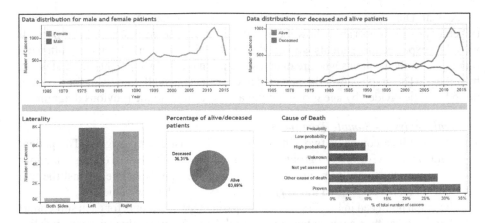

Fig. 5. SBCDS Dashboard example

Survival Analysis – Time and Age Groups

Analysing the survival time of patients is more complicated than creating the basic dashboard. First, before beginning the process, groups are needed to compare the survival time – e.g. using age groups and grouping the diagnosis time into decades. As a basic filter, only deceased patients and those who are diagnosed with primary breast cancer are counted for these diagrams, which results in 5,836 patient records. Depending on their age some of these patients are also filtered out. Using Tableau, Fig. 6 suggests that the trend of survival time for this cohort of patients has improved in the past four decades, both in the short term and (cumulatively) in the longer term.

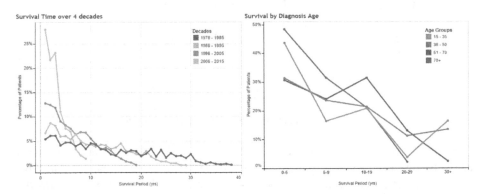

Fig. 6. Survival time comparision by decade and age group

Figure 6 [Left] illustrates how long patients have survived after breast cancer is initially diagnosed. One inference is that patients diagnosed later can survive longer. This is best seen from 0 to 10 years – after that many of the patients with a later diagnosed cancer are still alive. This is also shown in the relatively steep fall for the 2006–2015 decade, which corresponds to the newest diagnosed patients. These diagnoses are too recent to have a longer survival time.

Age_at_Diagnosis is one of the key sub-divisions of the data. Much is already known (whether proven or not) about how the age at diagnosis influences survival prospects. It would be expected that patients who are diagnosed with cancer at an early age often die more quickly than patients who are diagnosed later in life – Fig. 6 [Right] illustrates this hypothesis. The graph shows the percentage of patients overall who lived for a certain period of time divided by age group, giving a similar trend to the previous diagram.

Survival Analysis – Sample Treatments
The example here compares two different types of breast cancer surgery. The patient age-bands and survival period groups are defined using Tableau's grouping functionality. They are then used to produce visuals comparing survival time against the two different surgery types, with a much greater success than the raw values.

The graph produced in Fig. 7(a) shows the percentage of patients for each sub-cohort (age-band), rather than a percentage of the total. This illustrates that survival statistics for each surgery type can be compared appropriately, by looking at the percentage of patients in each age-band and survival period groups.

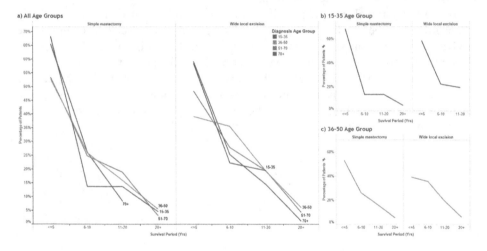

Fig. 7. Wide local excision vs simple mastectomy

Figure 7(b) shows the survival statistics for age-band 15–35. Both surgeries have a relatively high percentage of patients who survive less than five years. Figure 7(c) displays the same data but for age-band 36–50. The key difference here concerns the 6–10 years survival group, which has a greater percentage of patients following a wide local excision than a simple mastectomy.

Typically the graphs follow the common hypothesis that patients with cancer diagnosed at a young age have a reduced survival period. On the whole a greater percentage of patients are present in the 6–10, 11–20 and 20+ years groups for wide local excision, which could suggest that for this dataset the surgery is performing better than simple mastectomy. It is hard to derive any real conclusion from this analysis of course, but it does provide insight into which areas to concentrate further research.

4.2   Data Mining and Analytics

Sequential Patterns Mining

Querying the patient profiles is the starting point before pre-processing for sequential patterns mining. Based on the data model from Fig. 3, the relevant attributes selected for this purpose are: BCI_ID, Cancer_Category_Desc, Year_of_Death and Diagnosis_Month. Table 2 shows some raw output from an SQL query against the data extracted for the case study – the data has been limited to local referrals. Due to the limit of the table length, the query interleaves the date of each disease presentation event to give up to five event type/date pairs after the initial diagnosis.

Table 2. Sample results of patient event profiles

INITIAL	PRES1	PRES2	PRES3	PRES4	PRES5	STATUS
Primary	Loco-RR	Loco-RR	Risk-Reduce	-	-	Alive
Primary	Loco-RR	Other	Loco-RR	Metastatic	-	Dead
Primary	Loco-RR	Primary	Loco-RR	Loco-RR	Loco-RR	Dead
Primary	Metastatic	Metastatic	Metastatic	Metastatic	Metastatic	Dead
Primary	Other	Loco-RR	Loco-RR	Loco-RR	-	Alive
Primary	Primary	Metastatic	Other	Metastatic	Metastatic	Dead

Key: Primary Breast Cancer (Primary), Loco-Regional Recurrence (Loco-RR), Metastatic Disease (Metastatic), Risk Reducing Mastectomy (Risk-Reduce), Other Cancer Diagnoses (Other)

For the November 2015 dataset, there are 178 distinct disease event sequence profiles which correspond to 12,139 instances. The following pre-processing approach has been pursued to ensure the data is represented as accurately as possible: removal of instances where (1) there is no presentation at all; (2) initial presentation is anything other than primary breast cancer; (3) two or more presentations of primary cancer exist (when cancer is unilateral); and (4) the total number of events is less than three.

This dataset is then divided into two sub-groups: alive (188) and deceased (1,957). The GSP (Generalized Sequential Patterns) algorithm has been used through Weka for sequential patterns mining [3]. Five sequential patterns are shown below for alive patients under a minimum support threshold of *minsup* = 5%, where the numbers of patients are in brackets.

[1] <{Primary}{Loco-RR}{Loco-RR}> (39)
[2] <{Primary}{Loco-RR}{Metastatic}> (20)
[3] <{Primary}{Loco-RR}{Other}> (14)
[4] <{Primary}{Metastatic}{Metastatic}> (12)
[5] <{Primary}{Other}{Loco-RR}> (10)

These results are maximal patterns, i.e. they are not contained by other sequential patterns. When the same mining approach is applied to deceased patients and under the same setting, i.e. *minsup* = 5%, there are three sequential patterns:

<{Primary}{Loco-RR}{Loco-RR}{Metastatic}> (59)
<{Primary}{Loco-RR}{Metastatic}{Metastatic}> (55)
<{Primary}{Metastatic}{Metastatic}{Metastatic}> (49)

A directed acyclic Sequential Patterns Graph (SPG) has been used to represent the maximal sequence patterns [11]. Figure 8 shows the SPGs for both alive and deceased patients when $minsup = 5\%$. It can be seen that nodes of SPG correspond to elements (or disease events) in a sequential pattern and directed edges are used to denote the sequence relation between two elements. Any path from a start node to a final node corresponds to one maximal sequence and can be considered optimal.

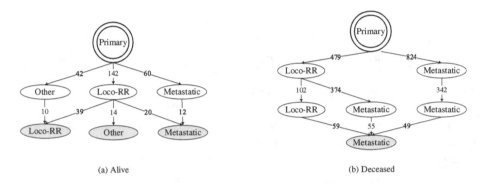

(a) Alive   (b) Deceased

Fig. 8. SPG for maximal sequential patterns when $minsup = 5\%$

Taking the left-side path from Fig. 8(a) as a sample for illustration, 42 patients are found with the diagnosis pattern of <{Primary}{Other}> with a support of at least 5% – out of this group there are another 10 presentations with Loco-Regional Recurrence. Down the left-side of Fig. 8(b), there are 479 patients with the diagnosis pattern of <{Primary}{Loco-RR}> – while 102 of these cases further present with Loco-Regional Recurrence, there are 374 instances of Metastatic Disease at the same level. These proceed respectively to another 59/55 presentations of Metastatic Disease.

Classification
All of the classification algorithms have been evaluated using Weka and compared according to a number of measures: (1) accuracy, sensitivity and specificity, (2) n-fold cross validation and (3) receiver operating characteristic. In this case study, a predictive model was created by applying the Weka Decision Tree J48 algorithm to the prepared dataset for the deceased patients whose cause of death was breast cancer on either the left or right side (i.e. not both). The 10-fold cross validation technique was used where the data was divided into 10 sub-sets of the same size.

The following variables have been selected for the predictor set: Max Tumour Grade, Positive Lymph Nodes and Age Group. The outcome status is either "Survive more than 5 years" or "Survive less than 5 years". There were 1,053 records containing full values for all the above variables. The complete decision tree is shown in Fig. 9 with 10 rectangular boxes representing the leaf nodes (i.e. classes). Each internal node represents a test on the variable and each branch represents an outcome for the test. There are two numbers in parentheses in the leaf nodes: the first shows how many patients reached this outcome and the second shows the number of patients for whom the outcome was not predicted to happen.

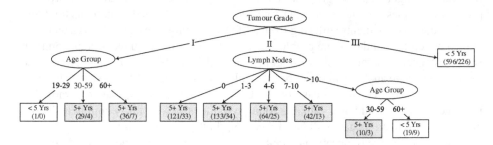

Fig. 9. Decision tree for survival analysis

Based on Max Tumour Grade – an indicator of how quickly a tumour is likely to grow and spread – some rules can be extracted from the decision tree in Fig. 9. For example (1) if the tumour is Grade III then the patient is likely to survive less than 5 years; (2) if Grade II and the number of positive lymph nodes is no more than 10, then the patient is likely to survive more than 5 years; and (3) if Grade I and the patient age group is '19–29', then they are likely to survive less than 5 years.

4.3 Discussion

It should be recognised that the modelling of cancer patient records at University Hospital Southampton is to a large extent feasible because of the adherence to a strict IT strategy for over 20 years. A fundamental challenge to analysing clinical data arises from the need to correctly and consistently identify the patient within the variety of systems which hold the data of interest. For a large organisation like UHS, which has a master patient index (MPI) containing well over 2 million individuals, correct identification against existing records is a constant challenge. UHS has pursued a range of approaches to achieving a high-quality MPI with low levels of duplicates and accurate patient details.

Another UHS/IT strategy that is pivotal to moving ahead with clinical research and data analysis is the relentless drive to concentrate data in relational databases, mainly Oracle. An integration engine (e.g. Ensemble) and HL7 messaging have also been used to facilitate high quality data exchange between the systems employed to deliver healthcare. Good quality data is acquired and maintained by good planning, prescribed and proven operational procedures, and professional IT development and systems management – good quality analytical work and sound conclusions can only come from such good quality data.

Temporally-structured clinical datasets are a relatively new tool in mainstream clinical practice. They pose significant challenges in data representation and analysis. The Southampton Breast Cancer Data System is a wholly new, very large and complex data presentation and analysis system which is in continual evolution and validation.

It provides an opportunity for experimentation with a range of data mining software tools and concepts, including the use of technology to identify outlier and "illogical" cohorts of patients with erroneous data.

Much thought must be given when deciding which data items to select for the proposed analyses and illustrations. Diagnostic tests are performed on patients for many reasons, and many patients suffering from a life threatening disease such as cancer may have other chronic and/or acute problems which almost certainly have little in common with the cancer. One must also consider that clinical data which could be of great value to the prospective analyses may not be available, either because it is not stored digitally or cannot be acquired. In the case of the UHS Lifelines concept, chemotherapy and radiotherapy data would be a valuable addition to the otherwise fairly complete dataset and it is hoped that this data will be forthcoming in the future.

Having undertaken appropriate pre-processing above, it was possible to perform sophisticated analyses on temporally-structured data using the software tools described. This work is iterative and with several objectives – one goal is to identify data mining tools which can be integrated into the bespoke data systems within the UHS-CDE. In time, this will allow clinicians more readily to analyse and understand the consequences of treatment decisions and their clinical outcomes across a spectrum of complex and chronic diseases.

Visual analytics is also becoming one of the essential techniques for health informatics. It allows users (clinicians, researchers, administrators and patients) to derive actionable and meaningful insights from vast and complex healthcare data. The use of software such as Tableau has the potential to improve graphical output from SBCDS, e.g. for histograms of year-on-year survival for any defined cohort. In addition, enhancement of timeline visualisation using the MVC architecture could provide an elegant solution which would allow UHS to move forward with Lifelines as a universal tool and testbed.

5 Conclusion

Within the healthcare domain there has been a vast amount of complex data generated and developed over the years through electronic patient records, disease diagnoses, hospital resources and medical devices. This data is itself a key resource and could play a vital role enabling support for decision making through processing information for knowledge extraction. The growing amount of data exceeds the ability of traditional methods for data analysis and this has led to the increasing use of emerging technologies. An overview process-driven framework has been proposed which integrates the preprocessing, data warehousing, data mining and visualisation aspects.

As part of the University Hospital Southampton clinical data environment, the Southampton Breast Cancer Data System has been developed as a "proof of concept" system. SBCDS is a unique timeline and episode-structured system for the analysis of the entire breast cancer pathway to final outcome of thousands of locally treated patients. There are already some valuable and complex analyses that have been developed within

SBCDS, and there is potential for further growth in functionality and capability of the system.

UHS Lifelines provides a conceptual model for the time-structured presentation of all key data for any patient or any chronic condition on a single computer screen. In particular, a Cancer-Lifetrak timeline has been developed to highlight the month of onset of key episodes of breast cancer progression, diagnosis, local recurrence, metastasis etc. It also permits measurement of time intervals between episodes and the correlation of these intervals with pathology and treatments.

One challenge of the Lifetrak representation is the overloading of the graphical interface by a concentration of many events over a relatively short period. A practical graphical user interface approach is thus needed which will handle this situation, so that the overriding story told by the data is not lost or corrupted. A sample dataset could be extracted for this purpose which includes the range of clinical data often associated with cancer patients, e.g. a mixture of pathology, radiology and clinical documentation events. The full set of records for all breast cancer patients comes to in excess of a million rows of event data and presents a bigger challenge for data management and predictive modelling.

Big data in healthcare is overwhelming not only because of its volume but also the diversity of data types and the speed at which it must be managed. The emerging NoSQL databases have significant advantages such as easy and automatic scaling, better performance and high availability. Using big data technologies has the potential to lead to more efficient and flexible healthcare applications. There are several challenges that need to be addressed to maximise the benefits of big data analytics in this area [24]: data quality; developing reliable inference methods to inform decisions; implementation of trusted research platforms; data analyst capacity; and clear communication of analytical results.

This paper has sought to give insight into how retrospective analysis of cancer treatment has the potential to identify some treatment pathways as more successful in terms of statistical outcome than others. The application of data mining and visual analytics could help prevent patients from being given sub-optimal treatment and help focus resources on treatments that are more successful. The cumulative benefits, both human and financial, might be enormous and it is hoped that others will take up the challenge to develop analyses where data is available to support that goal.

Acknowledgements. This research project has been supported in part by a Southampton Solent Research Innovation and Knowledge Exchange (RIKE) award for "Solent Health Informatics Partnership" (Project ID: 1326). The authors would like to thank Solent students who made some contribution to the work: in particular Chantel Biddle, Adam Kershaw and Alex Potter. We are also pleased to acknowledge the generous support of colleagues in the University Hospital Southampton Informatics Team, in particular Adrian Byrne and David Cable.

References

1. Bonadonna, G., Hortobagyi, G.N., Valagussa, P.: Textbook of Breast Cancer: A Clinical Guide to Therapy. CRC Press, Boca Raton (2006)
2. Devi, R.D.H., Deepika, P.: Performance comparison of various clustering techniques for diagnosis of breast cancer. In: IEEE International Conference on Computational Intelligence and Computing Research, pp. 1–5 (2015)
3. Hall, M., Frank, E., Holmes, G., Pfahringer, B., Reutemann, P., Witten, I.H.: The WEKA data mining software: an update. SIGKDD Explor. **11**(1), 10–18 (2009)
4. Han, J.W., Kamber, M., Pei, J.: Data Mining: Concepts and Techniques. Elsevier, New York (2011)
5. Hand, D.J., Smyth, P., Mannila, H.: Principles of Data Mining. MIT Press, Cambridge (2001)
6. Holzinger, A.: Trends in interactive knowledge discovery for personalized medicine: cognitive science meets machine learning. IEEE Intell. Inform. Bull. **15**(1), 6–14 (2014)
7. Hu, H., Correll, M., Kvecher, L., Osmond, M., Clark, J., et al.: DW4TR: a data warehouse for translational research. J. Biomed. Inform. **44**(6), 1004–1019 (2011)
8. Jerez-Aragones, J.M., Gomez-Ruiz, J.A., Ramos-Jimenez, G., et al.: A combined neural network and decision trees model for prognosis of breast cancer relapse. Artif. Intell. Med. **27**(1), 45–63 (2003)
9. Kimball, R., Ross, M.: The Data Warehouse Toolkit – The Definitive Guide to Dimensional Modeling. Wiley, New York (2013)
10. Lee, Y.J., Mangasarian, O.L., Wolberg, W.H.: Survival-time classification of breast cancer patients. Comput. Optim. Appl. **25**(1–3), 151–166 (2003)
11. Lu, J., Chen, W.R., Adjei, O., Keech, M.: Sequential patterns post-processing for structural relation patterns mining. Int. J. Data Warehouse. Min. **4**(3), 71–89. (2008). IGI Global, Hershey, Pennsylvania
12. Lu, J., Hales, A., Rew, D., Keech, M., Fröhlingsdorf, C., Mills-Mullett, A., Wette, C.: Data mining techniques in health informatics: a case study from breast cancer research. In: Renda, M.E., Bursa, M., Holzinger, A., Khuri, S. (eds.) ITBAM 2015. LNCS, vol. 9267, pp. 56–70. Springer, Cham (2015). doi:10.1007/978-3-319-22741-2_6
13. Lu, J., Hales, A., Rew, D., Keech, M.: Timeline and episode-structured clinical data: Pre-processing for data mining and analytics. In: 32nd IEEE International Conference on Data Engineering (ICDE) – Workshop on Health Data Management and Mining, pp. 64–67 (2016)
14. Mahajan, R., Shneiderman, B.: Visual and textual consistency checking tools for graphical user interfaces. IEEE Trans. Softw. Eng. **23**(11), 722–735 (1997)
15. Marr, B.: Big Data: Using Smart Big Data Analytics and Metrics to Make Better Decisions and Improve Performance. Wiley, Chichester (2015)
16. Martin, M.A., Meyricke, R., O'Neill, T., Roberts, S.: Mastectomy or breast conserving surgery? Factors affecting type of surgical treatment for breast cancer: A classification tree approach. BMC Cancer **6**, 98 (2006)
17. National Information Board. Personalised Health and Care 2020 (2014). https://www.gov.uk/government/uploads/system/uploads/attachment_data/file/384650/NIB_Report.pdf
18. NHS. Five year forward view (2014). http://www.england.nhs.uk/wp-content/uploads/2014/10/5yfv-web.pdf
19. Razavi, A.R., Gill, H., Ahlfeldt, H., Shahsavar, N.: Predicting metastasis in breast cancer: Comparing a decision tree with domain experts. J. Med. Syst. **31**, 263–273 (2007)
20. Reenskaug, T., Coplien, J.: The DCI architecture: A new vision of object-oriented programming (2009). http://www.artima.com/articles/dci_vision.html

21. Reps, J., Garibaldi, J.M., Aickelin, U., Soria, D., Gibson, J.E., Hubbard, R.B.: Discovering sequential patterns in a UK general practice database. In: IEEE-EMBS International Conference on Biomedical and Health Informatics, pp. 960–963 (2012)
22. Rew, D.: Issues in professional practice: The clinical informatics revolution. Published by Association of Surgeons of Great Britain and Ireland (2015)
23. Stolba, N., Tjoa, A.: The relevance of data warehousing and data mining in the field of evidence-based medicine to support healthcare decision making. Int. J. Comput. Syst. Sci. Eng. **3**(3), 143–148 (2006)
24. Wyatt, J.: Plenary Talk: Five big challenges for big health data. In: 8th IMA Conference on Quantitative Modelling in the Management of Health and Social Care (2016)

Poster Session

Contextual Decision Making
for Cancer Diagnosis

Samia Sbissi$^{(\boxtimes)}$ and Said Gattoufi

SMART, Higher Institute of Management of Tunis (ISG, Tunisia),
University of Tunis, Tunis, Tunisia
samia.sbissi@gmail.com, algattoufi@gmail.com

Abstract. Pathologist needs to routinely make management decisions about patients who are at risk for a disease such as cancer. Although, making a decision for cancer diagnosis is a dificult task since it context dependent. The term context contains a large number of elements that limits strongly any possibility to automatize this. The decision-making the process being highly contextual, the decision support system must benefit from its interaction with the expert to learn new practices by acquiring missing knowledge incrementally and learning new practices, it is called in deferent research human/doctor in the loop, and thus enriching its experience base.

Keywords: Decision making · Decision support system · Cancer diagnosis

1 Introduction

How doctors think, reason and make clinical decisions is arguably their most critical skill. Importantly, it underlies a major part of the process by which diagnoses are made. In all of the varied clinical domains where medicine is practiced, from the anatomic pathology laboratory to the intensive care unit, decision making is a critical activity. Medical diagnostic decision support systems have become an established component of medical technology. The main concept of the medical technology is an inductive engine that learns the characteristics of the diseases and can then be used to diagnose future patients with uncertain disease states. A cancer diagnosis is a human activity that is context-dependent. The context contains a large number of elements that limits strongly any possibility to automatize this. Recently, digitization of slides prompted pathologists to migrate from slide analysis under a microscope to slide image analysis on the screen. With the advancement in digital computing technology, many researchers have combined image processing and artificial intelligence to develop intelligent assistant systems to assist the pathologist in the diagnosis process. An Intelligent Assistant System (IAS) is an agent that is designed to assist an expert in his domain (in our case the expert is a pathologist). The goal of the intelligent assistant system is to help intelligently the expert in his decision-making process, not to provide

© Springer International Publishing AG 2017
M. Bursa et al. (Eds.): ITBAM 2017, LNCS 10443, pp. 55–65, 2017.
DOI: 10.1007/978-3-319-64265-9_5

an additional expertise in the domain. An IAS must be designed and developed in a formalism providing a uniform representation of knowledge, reasoning, and contextual elements.

The logic of the design and the realization of our system is built around three phases: (1) the capture of the information, (2) The structure of the information and (3) The content analysis,

1. Phase of capture: it is to capture and collect any information useful for the achievement of our process of modeling our consultant. There are different types of information:
   - Medical Information: This information concerns how the pathologists proceed in the diagnosis, it contains all the information of the expertise of the field.
   - Information for the support of reasoning: it represents all types of media for which we take the decision: example a medical imaging, medical record.
2. Phase of structure: it is to structure and organize the information captured; this phase is important because of this last that derives the heart of our consultant.
3. Analysis phase: it must enable users to take the right decision in any context.

The rest of the paper is organized as follows. Section 2 discusses the main idea of the domain and a brief summary of the decision analytic model. Section 3 represents a brief literature review on cancer diagnosis and modeling cancer diagnosis. Section 4 is a summary of computer aided diagnosis. Section 5 is concentrated on the context and his role in medical decision making. Section 6 describes our contribution and a critical discussion. Section 7 ends this article with a conclusion.

2   Model and Decision Analytic Model

A model represents a physical structure that have an actual entity or can be a series of mathematical equations that describe supply and demand.

A decision model often incorporates mathematical functions that represent what we know about the disease or the relationships between risk factors and disease, or disease and clinical outcomes. The impact of these assumptions can be evaluated in sensitivity analysis.

Different types of decision-analytic model structures are typically used in medical decision-making applications, but we propose to present briefly five models (decision tree, markov (cohort) model, microsimulation (individual) model, Dynamic model, Discrete event simulation model).Finally, we provide a summary of the different of models in Table 1.

Table 1. Summary of types of decision model structures

Model type	General description	Type of decision best suited for
Decision tree	Diagrams the risk of events and states of nature over a fixed time horizon	Interventions for which the relevant time horizon is short and fixed
Markov (cohort) model	Simulates a hypothetical cohort of individuals through a set of health states over time	Modeling interventions for diseases or conditions that involve risk over a long time horizon and/or recurrent events
Microsimulation (individual) model	Simulates one individual at a time; tracks the past health states of individual and models risk of future events stochastically	Modeling complex disease processes, when Markov models are too limiting
Dynamic model	System of differential equations that simulates the interactions between individuals and the spread of disease	Modeling interventions for communicable diseases, such as vaccinations
Discrete event simulation model	Simulates one individual at time as well as interactions among individuals or within a health care system	Evaluating alternative health care systems (e.g., workflow, staffing) though flexible enough to address questions in several different areas

3   Cancer Diagnosis and Modeling Cancer Diagnosis

3.1   Cancer and Cancer Diagnosis

Cancer is a disease in which a group of cells exhibits irregular cell growth cycle. In a normal cell cycle, the cells undergo mitosis process to replicate itself and hence the cell grows [13,14] The importance of making a timely and accurate diagnosis is fundamental for safe clinical practice and prevention error [6]. The task of making such a diagnosis involves doctors integrating key information from across all the stages of the clinical inquiry (including history taking, physical examination, and investigations).

Making an appropriate clinical diagnosis remains fundamentally important for the patient for the outcome initiates a cascade of subsequent actions, such as prescribing a drug or performing an operation, with real-world consequences.

Pathologists are problem-solvers, fascinated by the process of disease and eager to unlock medical mysteries, such as cancer, using the sophisticated tools and methods of modern laboratory science.

Histopathology is a branch of pathology that deals with the examination of tissue sections under a microscope in order to study the manifestation of diseases. The tissues are processed, sectioned, placed onto glass slides and are stained. The stained slides are examined under the microscope by a pathologist.

An important tool for the detection and management of cancer is an analysis of tissue samples under the microscope by a pathologist. Looking at the cancer cells under the microscope, the pathologist looks for certain features that can help predict how likely the cancer is to grow and spread. These features include the spatial arrangement of the cells, morphological characteristics of the nuclei, whether they form tubules, and how many of the cancer cells are in the process of dividing (mitotic count). These features have taken together, determine the extent or spread of cancer at the time of diagnosis.

3.2   Modeling Cancer Diagnosis

Cancer has been characterized as a heterogeneous disease consisting of many different subtypes. [12] the early diagnosis and prognosis of a cancer type have become a necessity in cancer research, as it can facilitate the subsequent clinical management of patients.

Machine learning is a branch of artificial intelligence that employs a variety of statistical, probabilistic and optimization techniques that allows computers to "learn" from past examples and to detect hard-to-discern patterns from large, noisy or complex data sets. This capability is particularly well-suited to medical applications, especially those that depend on complex proteomic and genomic measurements. As a result, machine learning is frequently used in cancer diagnosis and detection. [12]: a variety of techniques in Machine learning have been used as an aim to model the progression and treatment of cancerous conditions and is not new to cancer research, including Artificial Neural Networks (ANNs), Bayesian Networks (BNs), Support Vector Machines (SVMs) and Decision Trees (DTs) have been widely applied in cancer research for the development of predictive models, resulting in effective and accurate decision making. According to the latest PubMed statistics, more than 1500 papers have been published on the subject of machine learning and cancer. However, the vast majority of these papers are concerned with using machine learning methods to identify, classify, detect, or distinguish tumors and other malignancies.

The complexity of cancer development manifests itself on at least three scales that can be distinguished and described using mathematical models, namely, microscopic, mesoscopic and macroscopic scales. [11] proposes a neural-fuzzy approach for modeling breast cancer diagnosis. The neural-fuzzy approach models the diagnosis system as a three-layered neural network. The first layer represents input variables with various patient features, the hidden layer represents the fuzzy rules for diagnostic decision based on the input variables; and the third layer represents the output diagnostic recommendations. [20] describes the delays that occur in the detection, diagnosis, and treatment of cancers, including those of the head and neck. This model comprises four-time intervals that together make up the total time between the appearance of signs or symptoms of cancer and the commencement of treatment. These intervals are: appraisal, help seeking, diagnostic and pre-treatment.

The technical aspects of some of the ontology-based medical systems for cancer diseases is discusses in [1]. They proposed ontology based diagnostic methodology for cancer diseases. This methodology can be applied to help patients, students, and physicians to decide what cancer type the patient has, what is the stage of cancer and how it can be treated. The proposed methodology contains three basic modules namely, the diagnostic module, the staging module, and the treatment recommendation module. Each module (diagnostic, staging or treatment recommendation) can be applied as a stand-alone application according to the user requirements. For example, if the user knows the cancer type and wants to determine the cancer stage it can use the staging module by providing the cancer type and signs and symptoms and the module will determine the current stage and so on.

4   Computer Aided/Assisted Diagnosis (CAD)

Researchers in medical informatics are interested in using computers to assist physicians and other health care personnel in difficult medical decision-making tasks such as diagnosis, therapy selection, and therapy evaluation. Clinical decision-support systems are computer programs designed to help health care personnel in making clinical decisions Shortlifie et al. (1990) Since one of the first reported systems in 1964, the field has matured considerably and has produced systems for various medical domains. Notable among these are MYCIN for the selection of antibiotic therapy, INTERNIST-1 for diagnosis in general internal medicine, and ONCOCIN for the management of cancer patients.

Various signal processing and machine learning techniques have been introduced to perform computer-aided detection and diagnosis. Early works have focused on image processing and classification techniques to extract features of the image and predict the outcome (i.e., whether benign or malignant) in the image [8,19]. A neural network-based algorithm [2] and the use of an ensemble of SVMs aiming to transform the feature vector to a new representation vector [18] are proposed to solve the diagnosis problem.

Initially, it had been employed for radiology images [7], but over the last decade, it has also found its application with histopathology images. The area of Computer Aided Diagnosis combines methods or algorithms from Digital Image Processing and Pattern Recognition. It has been shown in a study that CAD has relieved the pathologist from the routine monotonous work of scanning slides [9].

5   Context and the Role of Context in Medical Decision Making

5.1   Context

The role of context can greatly influence how knowledge is organized and utilized in a knowledge-based system. Brezillon defines context from an engineer's position "as the collection of relevant conditions and surrounding influences that make a situation unique and comprehensible" [3]. This definition illustrates the potential value of the context for the purpose of providing focus on a system. The context allows groupings of knowledge for specific situations, thus the system can avoid superfluous questions unrelated to the current context.

At least, there is now a consensus around the following definition **context is what constrains reasoning without intervening in it explicitly** [3].

5.2   Modeling Context for Medical Decision Making

In order to integrate context in the process of diagnosis, [17] proposed a novel design framework for a computer-aided in breast cancer diagnosis system. The system incorporates contextual information and makes diagnostic recommendations to physicians, aiming to minimize the false positive rate of diagnosis, given a predefined false negative rate.

They considered the contextual information of the patient (also known as situational information) that affects diagnostic errors for breast cancer. The contextual information is captured as the current state of a patient, including demographics (age, disease history, etc.), the breast density, the assessment history, whether the opposite breast has been diagnosed with a mass, and the imaging modality that was used to provide the imaging data. The proposed algorithm is an on-line algorithm that allows the system to update the diagnosis strategy over time. The proposed algorithm is an on-line algorithm that allows the system to update the diagnostic strategy over time. They proposed a diagnostic recommendation algorithm that is formulated to make diagnostic recommendations over time. The algorithm exploits the dynamic nature of patient data to learn from and minimize the false positive rate of diagnosis given a false negative rate.

In [15], their goal is to achieve a semantic platform which is open and generic for digitized histology integrating a cognitive dimension. The system offers automatic semantic annotations to reach for supporting a diagnosis, taking into account the explicit and implicit medical knowledge field, reflections of the pathologist and contextual information of breast cancer gradation.

And this as part of a project "Mico": Cognitive microscope (Mico) that aims for a change in medical practices by providing an environment new medical imaging Anatomy and Cytology Pathological, enabling reliable decision making in histopathology.

[5], Author focuses on the formalization of medical practices in chronic inflammatory bowel disease diagnosis as a contextual graph to identify a consensus methodology. They have identified a "Glocal" search in the decision-making process with a global exploration for detecting zones of interest, and a zoom inside the zones of interest.

The term "Glocal" is proposed as an association of global and local. It associates a phase of superficial exploration for finding a zone of interest and a phase of fine-grained analysis of a zone of interest. This search concerns contextual elements at different levels of granularity as identified from the analysis of digital slides and they discussed the role of the local approach in other domains like Control and Command rooms for the army and subway-line monitoring.

For a continuing of the project "MICO" and in order to support the evolution towards digital pathology, the key concept of [16] approach is the role of the semantics as driver of the whole slide image analysis protocol. All the decisions being taken into a semantic and formal world.

As part of modeling decision making using contextual graphs, [4] present an example of a workflow manager that is developed in a large project in breast cancer diagnosis. The workflow manager is a real-time decision-making process in which the result of an action may change the context of the decision-making process. Their goal was developing a decision support system for users who have a high level of expertise in a domain not well known or too complex. Their expertise is composed of chunks of contextual knowledge built mainly by experience under a compiled expression. The decision-making process uses this expertise to propose critical and definitive decisions. In the MICO project, the expert is an anatomo-cyto-pathologist who analyzes digital slides (coming from biopsies) to diagnose if a patient in a surgery has or does not have breast cancer.

6 Critical Discussion and Our Approach

6.1 Discussion

The importance of making a timely and accurate diagnosis is fundamental for safe clinical practice and preventing error and the developers of new technologies conduct much research to develop a Decision Support System that is able of helping pathologist in making contextual decisions for cancer diagnosis. [17] discusses the performance of context for minimizing diagnostic errors by identifying what knowledge or information is most influential in determining the correct diagnostic action in order to minimize the false positive rate of the diagnosis is given a false negative rate, but they do not offer for pathologists actions in order to do to reach a correct diagnosis.

The role of context can greatly influence how knowledge is organized and utilized in a knowledge-based system that is proved in the majority of papers. Many technical were introduced to build Decision Support System for Cancer diagnosis, which integrates a numeric slide and others integrate clustering technique. But the problem that we do not find a search that integrates both the both technical. Another important point is that the majority of the proposed work treating breast cancer, but not rigorously examine other cancers.

The integration of the knowledge of a domain expert may sometimes greatly enhance the knowledge discovery process pipeline. The combination of both human intelligence and machine intelligence, by putting a "humanin-the-loop" would enable what neither a human nor a computer could do on their own. This human-in-the-loop can be beneficial in solving computationally hard problems, where human expertise can help to reduce an exponential search space through heuristic selection of samples, and what would otherwise be an NP-hard problem, reduces greatly in complexity through the input and the assistance of a medical doctor into the analytics process [10,21].

6.2 Our Contribution

A cancer diagnosis is a human activity and context-dependent. The term context contains a large number of elements that limits strongly any possibility to automatize and it's due that the medicine is an art, not an exact science and rational because:

1. It uses the biological knowledge to try to explain and understand the "etiology" of disorders observed in a person.
2. There are too many parameters to be considered and it's impossible to establish parameters of recognition because of the independence between parameters.
3. This discipline is also called the art of diagnosis. The known diseases diagnostics will eventually understand the etiology of a disease. Epidemiological data, statistics, and specific to the patient guide the therapeutic choice.
4. It has the role of ensuring the health of society. She must know cure, but also heal. It has a role in solving social problems and exclusion in the relief of pain problems and support people in later life. It must also have a preventive and educational role for the population.

The role of context can greatly influence how knowledge is organized and used in a knowledge-based system.

Our goal is to develop a decision support system to the pathologist that has a high level of expertise in their domain that analyses digital slides (coming from biopsies) to diagnose if a patient in a surgery has or not cancer.

The decision support system must:

1. behave as an intelligent assistant, following what the expert is doing, how he is doing it, anticipating potential needs.
2. work from the practices developed by the experts with all the contextual elements used by the expert during practice development.

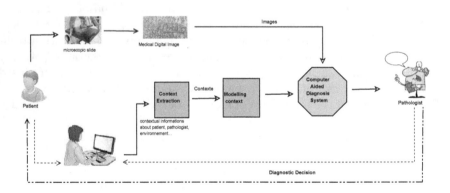

Fig. 1. Context informations

Figure 2 shows our approach how it is based on the context and we are going to explain the choice of different contexts which may exist. We will explain each contextual element by taking an example. In a service Anapath the analysis of a blade can be done in different ways: systematic (from left to right and from top to bottom: thus the direction of viewing) or well to do a search on areas of

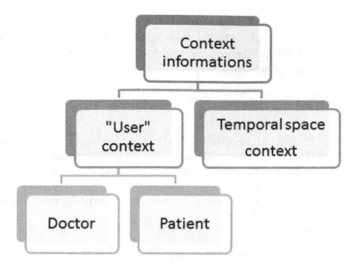

Fig. 2. Cancer diagnosis process overview

interest specific well or do a zoom in the first area encountered it is a way of research without any reflection, etc.

Other contextual elements may be taken into account example:

- the person who makes the analysis (doctor or technician), the conditions of work (at the beginning of the day or at the end of the day),
- the temperature may also have an influence on the decision-making example. In Anapath service, the levy is going to have a displacement of a lab to the other,
- a question must be asked "if the technician who place the levy decided to take a break or have a cigarette outside the service, is this the time spent and the climate change may not affect the temperature and the quality of sampling and by the suite on its viewing and interpretation by the doctors?

We distinguish also many other sources of contextual elements according to the doctor himself:

- Its preferences, point of view, etc.
- Also on the task to achieve (patient already come 3 times).
- On the situation where the task is performed (period of work overload, patient "recommended" by the boss, etc.)
- The local environment (the equipment just to have a fault).

We can conclude that all these contextual elements can have an influence on the quality of the diagnosis and also on the decision-making.

7   Conclusion

The diagnosis of cancer [13] is a human activity that depends on the context in which it is made. Scanning slides prompted pathologists to pass from analyzing glass slide to the screen. This migration offer the possibility of procedural on at least part of their analytical methods, also integrating other types of reasoning support offer a large flexibility and expand the reasoning field with the aim of improving the quality of service.

References

1. Alfonse, M., Aref, M.M., Salem, A.B.M.: An ontology-based cancer diseases diagnostic methodology. Recent Advances in Information Science. https://scholar. google.com/scholar?q=An+ontology-based+cancer+diseases+diag-+nostic+ methodology.+Recent+Advances+in+Information+Science&hl=fr&as_sdt=0& as_vis=1&oi=scholart&sa=X&ved=0ahUKEwi28d3pxYnVAhWEXRQKHcBVDq UQgQMIIzAA
2. Arajo, T., Aresta, G., Castro, E., Rouco, J., Aguiar, P., Eloy, C., Polnia, A., Campilho, A.: Classification of breast cancer histology images using convolutional neural networks. PLOS ONE **12**(6), 1–14 (2017)
3. Brézillon, P., Pomerol, J.: Contextual knowledge sharing and cooperation in intelligent assistant systems. Le Travail Humain - PUF **62**(3), 223–246 (1999)
4. Brézillon, P., Aroua, A.: Representation of real-time decision-making by contextual graphs based simulation. J. Decis. Syst. **22**(1), 28–42 (2013)
5. Brézillon, P., Attieh, E., Capron, F.: Modeling Global Search in a Decision-Making Process, pp. 80–91 (2014)
6. Croskerry, P.: The importance of cognitive errors in diagnosis and strategies to minimize them. Acad. Med. J. Assoc. Am. Med. Coll. **78**(8), 775–780 (2003)
7. Fujita, H., Uchiyama, Y., Nakagawa, T., Fukuoka, D., Hatanaka, Y., Hara, T., Lee, G.N., Hayashi, Y., Ikedo, Y., Gao, X., et al.: Computer-aided diagnosis: the emerging of three cad systems induced by Japanese health care needs. Comput. Methods Programs Biomed. **92**(3), 238–248 (2008)
8. Giger, M.: Computer-aided diagnosis in diagnostic mammography and multimodality breast imaging. RSNA Categorical Course Diagn. **60637**, 205–217 (2004)
9. Hipp, J., Flotte, T., Monaco, J., Cheng, J., Madabhushi, A., Yagi, Y., Rodriguez-Canales, J., Emmert-Buck, M., Dugan, M.C., Hewitt, S., et al.: Computer aided diagnostic tools aim to empower rather than replace pathologists: lessons learned from computational chess. J. Pathol. Inform. **2**(1), 25 (2011)
10. Holzinger, A.: Interactive machine learning for health informatics: when do we need the human-in-the-loop? Brain Inform. **3**(2), 119–131 (2016)
11. Keleş, A., Keleş, A., Yavuz, U.: Expert system based on neuro-fuzzy rules for diagnosis breast cancer. Expert Syst. Appl. **38**(5), 5719–5726 (2011)
12. Kourou, K., Exarchos, T.P., Exarchos, K.P., Karamouzis, M.V., Fotiadis, D.I.: Machine learning applications in cancer prognosis and prediction. Comput. Struct. Biotechnol. J. **13**, 8–17 (2015)
13. Lee, H., Chen, Y.P.P.: Cell cycle phase detection with cell deformation analysis. Expert Syst. Appl. **41**(6), 2644–2651 (2014)
14. Nahar, J., Tickle, K.S., Ali, A.B., Chen, Y.P.P.: Significant cancer prevention factor extraction: an association rule discovery approach. J. Med. Syst. **35**(3), 353–367 (2011)
15. Naour, G.L., Genestie, C., Roux, L., Veillard, A., Racoceanu, D., Capron, F.: Un explorateur visuel cognitif (MIcroscope COgnitif-MICO) pour l histopathologie. Application au diagnostic et à la graduation du cancer du sein, pp. 1–2 (2004)

16. Racoceanu, D., Capron, F.: Towards semantic-driven high-content image analysis: an operational instantiation for mitosis detection in digital histopathology. Comput. Med. Imaging Graph. **42**, 2–15 (2015)
17. Song, L., Hsu, W., Xu, J., Schaar, M.V.D.: Using contextual learning to improve diagnostic accuracy: application in breast cancer screening. IEEE **2194**(c), 1–12 (2014)
18. Tsochatzidis, L., Zagoris, K., Arikidis, N., Karahaliou, A., Costaridou, L., Pratikakis, I.: Computer-aided diagnosis of mammographic masses based on a supervised content-based image retrieval approach. Pattern Recogn. **71**, 106–117 (2017)
19. Urmaliya, A., Singhai, J.: Sequential minimal optimization for support vector machine with feature selection in breast cancer diagnosis. In: 2013 IEEE 2nd International Conference on Image Information Processing, IEEE ICIIP 2013, pp. 481–486 (2013)
20. Walter, F., Webster, A., Scott, S., Emery, J.: The andersen model of total patient delay: a systematic review of its application in cancer diagnosis. J. Health Serv. Res. Policy **17**(2), 110–118 (2012)
21. Wartner, S., Girardi, D., Wiesinger-Widi, M., Trenkler, J., Kleiser, R., Holzinger, A.: Ontology-guided principal component analysis: reaching the limits of the doctor-in-the-loop. In: Renda, M.E., Bursa, M., Holzinger, A., Khuri, S. (eds.) ITBAM 2016. LNCS, vol. 9832, pp. 22–33. Springer, Cham (2016). doi:10.1007/978-3-319-43949-5_2

A Review of Model Prediction in Diabetes and of Designing Glucose Regulators Based on Model Predictive Control for the Artificial Pancreas

Kyriaki Saiti[1]([envelope]), Martin Macaš[2], Kateřina Štechová[3], Pavlína Pithová[3], and Lenka Lhotská[2]

[1] Department of Cybernetics, Czech Technical University in Prague,
Prague, Czech Republic
krks1988@gmail.com
[2] Czech Institute of Informatics, Robotics and Cybernetics,
Czech Technical University in Prague, Prague, Czech Republic
[3] FN Motol University Hospital in Prague, Prague, Czech Republic

Abstract. The present work presents a comparative assessment of glucose prediction models for diabetic patients using data from sensors monitoring blood glucose concentration as well as data from *in silico* simulations. The models are based on neural networks and linear and nonlinear mathematical models evaluated for prediction horizons ranging from 5 to 120 min. Furthermore, the implementation of compartment models for simulation of absorption and elimination of insulin, caloric intake and information about physical activity is examined in combination with neural networks and mathematical models, respectively. This assessment also addresses the recent progress and challenges in designing glucose regulators based on model predictive control used as part of artificial pancreas devices for type 1 diabetic patients. The assessments include 24 papers in total, from 2006 to 2016, in order to investigate progress in blood glucose concentration prediction and in Artificial Pancreas devices for type 1 diabetic patients.

Keywords: Prediction · Diabetes · Model predictive control · Artificial pancreas

1 Introduction

Diabetes mellitus (or diabetes) is a chronic, lifelong condition that affects the body's ability to use the energy found in food. There are four major types of diabetes: type 1, type 2, other specific forms of diabetes and gestational diabetes. Type 2 diabetes is an autoimmune condition, caused by the body attacking its own pancreas with autoantibodies. For people with type 1 diabetes, the damaged pancreas does not make insulin. This type of diabetes is a polygenetic multifactorial disease for people carrying certain genetic predisposition in which their

© Springer International Publishing AG 2017
M. Bursa et al. (Eds.): ITBAM 2017, LNCS 10443, pp. 66–81, 2017.
DOI: 10.1007/978-3-319-64265-9_6

immune cells destroy pancreatic beta cells which produce insulin. Triggers for this autoimmune insulitis are still unknown with various trigger such as inflection or stress under investigation. Type 2 diabetes also referred to as adult-onset diabetes, but with the epidemic of obese and overweight kids, more teenagers are now developing type 2 diabetes. Type 2 diabetes was also called non-insulin-dependent diabetes. Insulin resistance is occurred before type 2 diabetes which leads insulin production to reduction.

The prevalence of diabetes is increasing worldwide, with many now referring to it as an epidemic. The scientific community focuses now on creating Clinical Decision Support Systems for supporting physicians and diabetics in order to prevent or postpone complications related to diabetes. The effectiveness of systems depends mostly on accurate short- and long- term model prediction of glucose levels. There have been a growing number of studies and programs in the past 10 years attempting to create prediction models and test their ability to make accurate predictions.

This paper will review the literature on prediction models derived from clinical data or *in silico* evaluation and will also address the design of glucose regulators based on Model Predictive Control. The review is not meant to be exhaustive but rather it provides an overview of this area, highlighting examples of promising approaches, outlining key challenges, and identifying needs for future research and practice.

2   Datasets

The quality of data from which a model is identified directly influences the predictive capability of the model. Input variables such as the target group of a study, the type of diabetes and the observation period are crucial issues in every project. Carbohydrates intake (usually measured in grams), exercise data and stress information are found to play a vital role in blood glucose concentration and can lead to hyper- or hypo- glycemia episodes or under specific conditions (high level of exercise, balanced nutrition) can stabilize the blood glucose levels.

In most of studies, patients suggested to use a Continuous Glucose Monitoring device for a specific period of time (ranging from 3 to 77 days), record meals and exercise in a diary and to use SenseWear Armband for recording energy expenditure of daily physical activities or exercise events [8,9,22,24]. Table 1 summarizes information about variables that were used in each study as well as duration of the observation.

Mougiakakou et al. [14], notably, used data from four children (15 to 22 years old), two female and two male with diabetes duration ranged from 6 to 13 years. Children have been monitored for a period between 3 to 5 days. For this period, patients recorded the time of insulin injection, the insulin type and dose as well as the carbohydrate intake.

Zhao et al. [25] performed, in addition to using clinical data, *in silico* evaluation for 10 adult subjects who were simulated using the Food and Drug Administration- accepted University of Virginia/University of Pandova metabolic simulator.

Finan et al. [6] generated a nominal dataset in which three simultaneous boluses and meals in the insulin-to-carbohydrate ratio were administered to a virtual subject to simulate a 20 g carbohydrate breakfast, 40 g lunch, and 60 g dinner over a period of 24 h. All datasets simulated 24 h of 5 min data with breakfast, lunch, and dinner administered at 8:00 am, 12:00 pm, and 6:00 pm, respectively. Four additional datasets were generated in which different levels of input excitation were achieved by varying the insulin-to-carbohydrate ratio for the meals. Five more datasets were then simulated which were identical to the first five except that the lunch bolus was halved. Thus, the subject underbolused for lunch and became mildly hyperglycemic. A correction bolus was then taken 90 min after lunch in order to return the subject to normoglycemia before dinner. In all datasets, the virtual subject's glucose levels were kept between 65 and 200 mg/dl, a clinically acceptable glucose range. Gaussian white noise was added to the simulated glucose values for all datasets.

Rollins et al. [15] used data consists of glucose concentration and physiological signals collected from 5 subjects with type 2 diabetes under free-living conditions (approximately, 23 days of data per subject). A Continuous Glucose Monitoring device was used for measuring the blood glucose levels every 5 min as well as the body monitoring system SenseWear Pro3 for collecting metabolic, physical activity and lifestyle information.

Table 1 summarizes the researches on glucose prediction, the diabetes type, the input variables and the observation period. CGM refers to Continuous

Table 1. Summary of works on glucose prediction and datasets ([a]: clinical study, [b]: *in silico*, [c]: clinical study-children, CGM: Continuous Glucose Monitoring Data, BG: Blood Glucose Levels, CHO: Carbohydrates, I: Insulin, S: Stress, PA: Physical Activity)

Study (Year)	Diabetes Type (No. of patients)	Input variables	Observation period
Mougiakakou et al. (2006) [14]	Type 1 (4[c])	CGM, CHO	3–5 days
Sparacino et al. (2007) [20]	Type 1 (28[a])	BG, I	48 h
Baghdadi et al. (2007) [17]	Type 1 (1[a])	BG, I, CHO, PA, S	77 days
Pappada et al. (2008) [16]	Type 1 (18[a])	CGM, CHO, PA	3–9 days
Zainuddin et al. (2009) [23]	Type 1 (1[a])	BG, I, CHO, PA, S	77 days
Stahl et al. (2009) [21]	Type 1 (1[a])	BG, I	6 months
Gani et al. (2009) [7]	Type 1 (9[a])	CGM	5 days
Valletta et al. (2009) [22]	Type 1 (18[a])	CGM, CHO, PA	25 days
Georga et al. (2011) [9]	Type 1 (7[a])	CGM, CHO, PA	5–14 days
Rollins et al. (2012) [15]	Type 2 (1[b])	CGM, CHO, PA	multiple days
Zhao et al. (2012) [25]	Type 1 (7[a]) (10[b])	BG, I, CHO	multiple days
Georga et al. (2013) [8]	Type 1 (27[a])	CGM, CHO, PA	5–22 days
Zarkogianni et al. (2015) [24]	Type 1 (10[a])	CGM, PA	6 days
Qadah et al. (2016) [4]	Type 1 (10[a])	CGM, CHO	1–60 days

Glucose Monitoring device; BG, Blood Glucose levels and CHO, carbohydrates (measured mostly in grams).

It is remarkable that the only works in which stress data were used as input to the models were [17, 23]. Patients' psychological conditions were supposed to affect blood glucose concentration as well as metabolic rate, although it is difficult to record such factors from subjects and also difficult to be used from the simulation and prediction models.

3   Evaluation Criteria

The predicted performance of these models have been evaluated according to different prediction horizons using both mathematical and clinical evaluation criteria. The most commonly used mathematical criterion is Root Mean Square Error and the most common calculated clinical criterion was the Clarke's Error Grid Analysis [2].

Root Mean Square Error is the standard deviation of residuals (i.e. prediction errors). Residuals are a measure of how far from the regression line data points are; Root Mean Square Error measures how spread out these residuals are. It shows how concentrated the data is around the line of best fit. Root mean square error is commonly used forecasting and regression analysis to verify experimental results.

The Clarke's Error Grid Analysis was developed in 1987 to quantify the clinical accuracy of patient estimates of their current blood glucose as compared to the blood glucose values obtained by their meters. Clarke et al. differentiated five zones in the graph (A to E), with the following meanings [2]:

- Zone A represents glucose values that deviate from reference values by 20% or less and those that are in the hypoglycemic range (<70 mg/dl), not only the predicted value but also the reference value. Those values are clinically exact and acceptable and for the clinical treatment will be correct.
- Zone B represents the glucose values that deviate from the reference values by more than 20%. In this zone we are close to unacceptable errors but the clinical treatment has a high probability of being correct. The values that fall within zone B are also clinically acceptable.
- The values included in Zone C-E are potentially dangerous, since the measure or prediction is far from being acceptable and the indicated treatment will be different from the correct. There is a high possibility of making clinically significant mistakes for values within this zones.

Additionally, Relative Absolute Deviation and Sum of Squares of the Glucose Prediction Error have been used to describe the deviation of the predicted values from the actual data. Moreover, Absolute Difference percent and the Mean Absolute Difference percent have been also performed for neural networks as well as median Relative Absolute Difference percentage.

Finally, FIT have been used for measuring the *goodness* of the model as well as Correlation Coefficient percentage for quantifying the type of correlation and dependence between two or more values. Mean Square Prediction Error and Energy of the Second-Order Differences were calculated in terms of matching the predicted values with the original ones. Table 2 matches the evaluation criteria with the works in which they have been used.

Table 2. Summary of works on glucose prediction and mathematical criteria

Study	Mathematical/Clinical criteria	Abbreviation
Mougiakakou et al.	Root Mean Square Error (mg/dl)	RMSE (mg/dl)
	Correlation Coefficient (%)	CC (%)
Sparacino et al.	Mean Square Prediction Error	MSPE
	Energy of the Second-Order Differences	ESOD
Baghdadi et al.	Root Mean Square Error (mmol/l)	RMSE (mmol/l)
Pappada et al.	Mean Absolute Difference (%)	MAD (%)
	Absolute Difference (%)	AD (%)
Zainuddin et al.	Root Mean Square Error (mmol/l)	RMSE (mmol/l)
Stahl et al.	Root Mean Square Error (mg/dl)	RMSE (mmol/l)
Gani et al.	Goodness of the model (%)	FIT (%)
Valletta et al.		
Georga et a.l	Root Mean Square Error (mg/dl)	RMSE (mg/dl)
	Error Grid Analysis (%)	EGA (%)
Rollins et al.	Median Relative Absolute Difference (%)	RAD (%)
	Sum of Squares of the Glucose Prediction Error (%)	SSGPE (%)
Zhao et al.	Root Mean Square Error (mg/dl)	RMSE (mg/dl)
	Error Grid Analysis (%)	EGA (%)
Georga et al.	Root Mean Square Error (mg/dl)-(mmol/l)	RMSE (mg/dl)-(mmol/l)
	Error Grid Analysis (%)	EGA (%)
Zarkogianni et al.	Root Mean Square Error (mg/dl)	RMSE (mg/dl)
	Correlation Coefficient (%)	CC (%)
	Error Grid Analysis (%)	EGA (%)
	Mean Absolute Relative Difference (%)	MAD (%)
Qadah et al.	Root Mean Square Error (mg/dl)	Root Mean Square Error (mg/dl)

4   Prediction Models

A model-based controller for an artificial pancreas has the potential to automatically regulate blood glucose levels based on available glucose measurements, insulin infusion, meal information and model predictions of future glucose trends. Thus the identification of simple, accurate glucose prediction models is a key step in the development of an effective artificial pancreas. Many empirical (or *data-driven*) modeling techniques have been evaluated in both *in silico* and clinical studies. The existing dynamic empirical models for type 1 diabetes include linear input-output models and nonlinear models such as neural networks. The linear models that have received the most attention for type 1 diabetes applications are autoregressive models and autoregressive with exogenous inputs models.

4.1   Linear/Nonlinear Dynamic Models

Zarkogianni et al. [24] used a Linear Regression model in order to compare results with neural networks. Coefficients of the model were calculated by applying the least squares method.

Zhao et al. [25] employed a Latent Variable-based technique to develop an empirical glucose prediction model from type 1 diabetes subject data. This modeling technique consisted of two steps. First, a Partial Least Square model was developed to predict future glucose concentrations (i.e. model output). In the second step, the Partial Least Square model was improved by postprocessing using Canonical Correlation Analysis.

Final et al. [6] implemented a family of linear models including an Autoregressive model, an Autoregressive with exogenous inputs model, an Autoregressive Moving Average with exogenous inputs model and a Box-Jenkins model.

Georga et al. [9] proposed a method comprised Compartment Models, known from pharmacokinetics, as well as Support Vector Machines were employed in order to provide individualized glucose predictions. Compartment models were used for simulating the pharmacodynamics of the insulin, the meal absorption and elimination and the impact of exercise on glucose-insulin metabolism.

Gani et al. [7] implemented an Autoregressive model to predict near-future glucose concentrations with acceptable time lags. In this study, three possible scenarios were compared and contrasted for near-future predictions; scenario I used raw glucose data to obtain unregularized Autoregressive model coefficients, scenario II employed smoothed glucose data to compute unregularized Autoregressive model coefficients and scenario III employed smoothed data to generate regularized Autoregressive model coefficients.

Sparacino et al. [20] implemented an first-order Autoregressive model and compared the results with a first-order Polynomial model. For both methods, at each sampling time, a new set of model parameters is first identified by means of weighted least squares techniques.

Valletta et al. [22] evaluated a Compartment Model for simulating insulin pharmacodynamics and a Compartment Model for simulating the meal absorption and elimination. Additionally, Gaussian Processes were suggested to model the glucose excursions in response to physical activity data in order to study its effect on glycaemic control.

Rollins et al. [15] proposed an univariate model which was developed using only CGM device data and a multivariate model using CGM data as well as metabolic, physical activity and lifestyle information from a multi-sensor armband.

Stahl et al. [21] implemented and compared an Autoregressive moving average with exogenous inputs with a Nonlinear Autoregressive moving average with exogenous inputs. In this study, Compartment Models were also developed for insulin kinetics simulation and food digestion.

4.2   Neural Network Models

Zarkogianni et al. [24] presented a Feedforward neural network, a Self-Organized Map and a Neuro-Fuzzy network. In this work, the mini-batch method was used, which is a hybrid method combining the batch and online method for training the neural networks.

Mougiakakou et al. [14] proposed a Feed-Forward neural network as well as a Recurrent neural network combined with an one-line Real Time Recurrent Learning algorithm in order to update the Recurrent neural network weights online. Moreover, in this paper, Compartment Models have been evaluated for insulin kinetics simulation (one model for each type of insulin) in combination with Compartment Model for food digestion.

Pappada et al. [16] implemented neural network models using NeuroSolution software (Neurodimension, Gainesville, FL). These neural networks were configured to forecast future blood glucose levels within a certain predefined time frame or predictive window.

Qadah et al. [4] proposed a special type of artificial neural network, the Time Sensitive artificial neural network in combination with two compartment models, one for simulating insulin kinetics and one meal absorption and elimination. The proposed prediction method are divided in two steps; In the first step, a Time Sensitive artificial neural network was utilized to predict subcutaneous glucose measurements at several points in the future. In the second step these values were used to predict the occurrence of hypoglycemia events within the given prediction horizon.

Zainuddin et al. [23] implemented an expert system based on Principal Component Analysis and Wavelet neural network for blood glucose model prediction. The proposed Wavelet neural network developed from different wavelet families in the hidden layer such as Mexican Hat, Gaussian wavelet and Morlet.

Bghdadi et al. [17] developed a Radial-Basis Function network involving three entirely different layers for four intervals (morning, afternoon, evening and night). The input layer is made up of source nodes. The second layer is a hidden layer of high enough dimension and the third layer is the output layer which supplies the response of the network to the activation patterns applied to the input layer.

5   Results

Most of the reviewed here focused on performing, testing and presenting results for specific prediction horizons (in minutes) and infinite trials. Table 3 summarizes results derived from ten studies [4, 6–9, 16, 24, 25]. The most preferable prediction horizons are 30-, 60-, 120- minutes; The commonly used evaluation criteria were the Root Mean Square Error and Clarke's Error Grid Analysis. As an example, for 120 min ahead prediction the most satisfactory Root Mean Square Error result performed by Georga et al. [8] in which the calculated RMSE was equal to 33.04 mg/dl. At the same prediction horizon, the best perform of

Clarke's Error Grid Analysis was conducted by Zarkogianni et al. [24] in which $EGA_{(A)}$ was equal to 88.86%.

Mougiakakou et al. [14] calculated the Root Mean Square Error separately for each patient and the best results were measured for the second patient (15 years old, female, 13 years duration of diabetes) where RMSE = 7.19 mg/dl for the Feed Forward neural network and RMSE = 11.58 mg/dl for the Recurrent neural network.

Zainuddin et al. [23] performed tests for four intervals; morning, afternoon, evening and night for three different wavelet families. The most reliable results were achieved with Gaussian Wavelet (RMSE = 0.0348 mmol/l (afternoon), RSME = 0.0330 mmol/l (evening) and RMSE = 0.0170 mmol/l (night)) and from Mexican Hat (RMSE = 0.0460 mmol/l (morning)).

Baghdadi et al. [17] also performed tests for four intervals; morning, afternoon, evening and night. The best results appeared in estimates for night interval where RMSE (training data) = 0, RMSE (validation data) = 0.0119, RMSE (test data) = 0.0118.

Stahl et al. [21] represented results for 15 min, 60 min, 120 min and for infinite horizon. The fit between the two signals is measured and for each prediction horizon the fit was equal to 99.97%, 93.99%, 60.83% and 11.41%, respectively.

6   Artificial Pancreas - Model Predictive Control

The artificial pancreas is a technology in development to support people with diabetes automatically control their blood glucose level by providing the substitute endocrine functionality of a healthy pancreas. There are several important exocrine (digestive) and endocrine (hormonal) functions of the pancreas, but it is the lack of insulin production which is the motivation to develop a substitute. While the current state of insulin replacement therapy is appreciated for its life-saving capability, the task of manually managing the blood sugar level with insulin alone is arduous and inadequate. The goals of the artificial pancreas are to improve insulin replacement therapy until glycemic control is practically normal as evident by the avoidance of the complications of hyperglycemia, and to ease the burden of therapy for the insulin-dependent diabetes. One approach to those aims is the medical equipment approach in which an insulin pump is used under closed loop control using real-time data from a continuous blood glucose sensor [1]. Figure 1 shows the steps and the improvement that have been done until 2010 in artificial pancreas.

The pursuit of a closed-loop artificial pancreas that automatically controls blood glucose for individuals with type 1 diabetes has intensified during the past decade. Continuous insulin infusion pumps have been widely available for over 30 years, but *smart pump* technology has made such devices easier to use and more powerful. Continuous Glucose Monitoring technology has improved and the devices are more widely available. A number of approaches are currently under study for fully closed-loop systems; most manipulate only insulin, while others manipulate insulin and glucagon. Algorithms include on-off (for prevention

Table 3. Summary of works on glucose prediction and their results ($EGA_{(A)}$: percentage of data which fell in Zone A (clinically accurate))

Evaluation criteria	Model	PH = 5	PH = 15	PH = 30	PH = 45	PH = 50	PH = 60	PH = 75	PH = 100	PH = 120	PH = 150	PH = 180	PH = infinite
RMSE	FNN [24]					25.70			38.51				
	SOM [24]		12.29			21.06			33.68				
	WFNN [24]		15.64			25.5			40.81				
	LRM [24]		15.51			26.39			44.52				
	LVX [25]	11.1	18.7	24.4		29.2							
	ARX [25]	11.3	19.5	25.5		30.3							
	LV [25]	11.3	19.7	26.0		31.2							
	SVR [9]	9.51	16.02			24.81			36.15				
	NARX [4]		13.29										
	SVR [8]	9.05	15.29			24.19			33.04				
	AR [7]		1.8			10.8		21.6					
CC	FNN [24]		95.42			90.80			79.12				
	SOM [24]		97.92			94.00			84.22				
	WFNN [24]		96.87			91.72			79.06				
	LRM [24]		96.81			91.22			77.99				
$EGA_{(A)}$	FNN [24]		88.46			88.83			88.18				
	SOM [24]		91.86			90.45			88.86				
	WFNN [24]		89.45			88.89			87.34				
	LRM [24]		87.57			87.17			84.61				
	LVX [25]	96.8	86.1	78.8		72.1							
	ARX [25]	96.5	84.9	77.5		70.1							
	LV [25]	96.4	84.9	76.3		68.4							
	ARX [6]	100				89			70			38	
	ARMAX [6]	100				99			94			81	
	BJ [6]	99				97			92			73	
	SVR [9]	98.58	92.54			79.96			60.10				
	SVR [8]	95.85	95.51			94.35			93.03				

(continued)

Table 1. (*continued*)

Evaluation criteria	Model	PH = 5	PH = 15	PH = 30	PH = 45	PH = 50	PH = 60	PH = 75	PH = 100	PH = 120	PH = 150	PH = 180	PH = infinite
MARD	FNN [24]		7.25				11.44		18.68				
	SOM [24]		5.34				9.36		15.99				
	WFNN [24]		7.38				12.38		20.42				
	LRM [24]		7.47				12.79		22.27				
MAD	NN [16]				6.7		8.9	11.7	14.5	16.6	18.9		
FIT	ARX [6]	86					51		25			22	
	ARMAX [6]	85					77		65			49	
	BJ [6]	84					73		59			36	
RAD	ARX [6]	2.8					7.6		12.9			23.8	
	ARMAX [6]	2.7					4.7		6.3			9.2	
	BJ [6]	3.0					5.2		7.4			11.4	
	ARMA [15]		4.24										
SSGPE	ARMA [15]		7.43										
MSPE	AR [20]		336	1218									
ESOD	AR [20]		85684	228895									

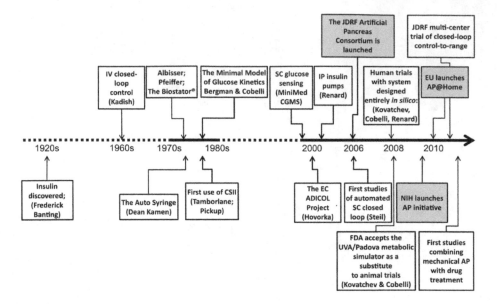

Fig. 1. Key milestones in the timeline of Artificial Pancreas progress [1].

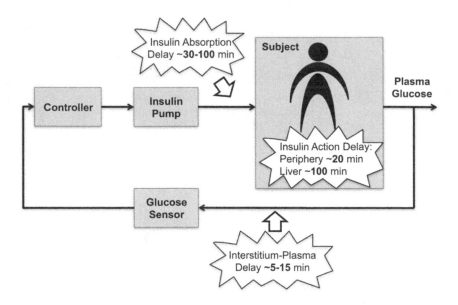

Fig. 2. Block diagram of closed-loop glucose control. Three major delays are indicated: insulin absorption (regular and ultrafast insulin), insulin action on peripheral tissues and on the liver, and sensing in the interstitium [1].

of overnight hypoglycemia), Proportional Integral Derivative, Model Predictive Control and Fuzzy Logic Based Learning Control. Meals cause a major *disturbance* to blood glucose, and in most causes mathematical models, such compartment models, are implemented. Model Predictive Control is a basic framework or strategy that can involve many different types of models and objective functions. Despite important developments in sensor and pump technology, the Artificial Pancreas must cope with the delays and inaccuracies in both glucose sensing and insulin delivery described in the previous sections. This is particularly difficult when a system disturbance, e.g., a meal, occurs and triggers a rapid glucose rise that is substantially faster than the time needed for insulin absorption and action (Fig. 2).

The new wave of control designs, Model Predictive Control, is based on prediction of glucose dynamics using a model of the patient metabolic system and, as a result, appears better suited for mitigation of time delays due to subcutaneous glucose sensing and insulin infusion. In addition, Model Predictive Control is a better platform for incorporation of predictions of the effects of meals and for introduction of constraints on insulin delivery rate and glucose values that safeguard against insulin overdose or extreme blood glucose fluctuations. In some sense, an Model Predictive Control algorithm works as a chess strategy [1]. On the basis of past game (glucose) history, a several-moves-ahead strategy (insulin infusion rate) is planned, but only the first move (e.g., the next 15-min insulin infusion) is implemented; after the response of the opponent, the strategy is reassessed, but only the second move (the 30-min insulin infusion rate) is implemented, and so on. In reality glucose prediction may be different from the actual glucose measurement or an unexpected event may happen; with this strategy these events are taken into account in the next plan.

Magni et al. [12] present an unconstrained Model Predictive Control strategy, where the model is a linearization of a nonlinear model, obtained at an average value of the population parameters. In simulation studies with a sample time of 30 min, they show that a single parameter, the weighting on the output predictions in the objective function, can be tuned for each individual for better performance.

Soru et al. [19] discuss techniques for meal compensation and individualization for better performance, in simulation studies involving four different scenarios and 100 subjects. They first use a single adjustable parameter based on clinical parameters. They then develop low-order models to produce more a more realistic model as a basis for Model Predictive Control; this model is further revised based on patient specific information. The Model Predictive Control algorithm in Soru et al. and the Hovorka et al. [10] MPC algorithm were studied in trials involving 47 patients in six centers, as reported by Devries et al. [2]; while the closed-loop algorithms each had a higher mean glucose than open-loop control, both resulted in less time in hypoglycemia than open-loop control.

An adaptive Generalized Predictive Control approach (based on recursive identification of Autoregressive with exogenous inputs models) is taken by Khatib et al. [3]. Lee et al. [11] use subspace identification techniques to develop discrete state space models, and incorporate constraints in Model Predictive Control; additional features include a pump shut-off algorithm to avoid hypoglycemia, and meal detection and meal size estimation algorithms to handle unan- nounced meals.

Markakis et al. [13] presents a study of the efficacy of regulating blood glucose in type 1 diabetics with a Model Predictive Control strategy, utilizing a nonparametric/Principal Dynamic Modes model. For this purpose, a stochastic glucose disturbance signal is introduced and a simple methodology for predicting its future values is developed. The results of our simulations confirm that the proposed algorithm achieves very good performance, is computationally efficient and avoids hypoglycaemic events.

More recently, Favero et al. [5] proposes a new modular Model Predictive Control algorithm focusing on meal control. Six type 1 diabetes mellitus patients underwent 42 h experiments: sensor−augmented pump therapy in the first 14 h (open−loop) and closed−loop in the remaining 28 h. Seron et al. [18] present a methodology for comparing benefits of regular Model Predictive Control versus stochastic Model Predictive Control especially when applied to type 1 diabetes patients. This work describes the use of a model which captures the relevant uncertainty associated with measurements and future disturbances, then one can use Monte Carlo simulation to obtain detailed information about the potential performance of an algorithm. The ideas were illustrated by computing the probability distribution of blood glucose levels for 4 different feedback control algorithms.

7   Discussion

The present paper reviewed fifteen studies about model prediction and nine works on Artificial Pancreas and the use of Model Predictive Control. The prediction models are strictly related to the process of designing Artificial Pancreas devices for controlling the blood glucose levels, preventing hypo- and hyperglycemia episodes and keeping levels near to euglycemia levels.

The quality and variety of data play a major role in the effectiveness of a model. First, researchers must consider to use clinical data or make *in silico* evaluations. The methods of data preprocessing as well as the implementation of compartment models for simulations are the second step for almostevery research reviewed here. As a result, the first part of this paper focused on clarifying the type of data used for each model.

Some researchers implemented linear and nonlinear mathematical models (AR, ARX, ARMA, NARMAX, SVR) while some others implemented neural networks (FNN, RNN, WFNN). According to the results, none of the models can yet be accepted as the ideal approach and this illustrates the need for further investigations and creations of more complex models.

Concerning the design of Artificial Pancreas devices, it is clear that a successful closed-loop system has a number of components, including sensors (in most cases a CGM device), actuators and prediction algorithms. Simulation studies have provided important results that enable fewer clinical tries, particularly for full closed-loop systems with given performance goals. Recent clinical results are very promising and a substantial number of out-patient trials are proceeding.

8   Conclusion

The nature of data (quality and variety) is crucially in each survey which is relevant to model prediction in diabetes. Several devices record blood glucose levels and insulin doses are used and there are many approaches to optimizing their implementation. The clinical studies focused on recording carbohydrates intake, physical activities and some recorded the psychological state of patients. It is clear that models developed with exogenous inputs are more reliable than simpler models while the combination of prediction models with compartment models provide better estimations for short term predictions.

Up to now, none of the models are considered capable of performing accurate long-term predictions. As a result, future work on prediction models for diabetes should focus on investigating and implementing linear/nonlinear models or neural networks, performing results for prediction horizons of more that two hours and comparing the results with those from short-term predictions.

Acknowledgements. Research has been supported by the AZV MZ CR project [No. 15-25710A] "Individual dynamics of glycaemia excursions identification in diabetic patients to improve self managing procedures influencing insulin dosage" and by CVUT institutional resources (SGS grant application [No. OHK-4-/3T/37]).

References

1. Cobelli, C., Renard, E., Kovatchev, B.: Artificial pancreas: past, present, future. Diabetes **60**, 2672–2682 (2011)
2. Clarke, W.L., Anderson, S., Farhy, L., Breton, M., Gonder-Frederick, L., Daniel, C., Boris, K.: Evaluating the clinical accuracy of two continuous glucose sensors using continuous glucose-error grid analysis. Diabetes Care **28**(10), 2412–2417 (2005)
3. El-Khatib, F.H., Jiang, J., Damiano, E.R.: Adaptive closed-loop control provides blood-glucose regulation using dual subcutaneous insulin and glucagon infusion in diabetic swine. J. Diabetes Sci. Technol. **1**(2), 181–192 (2007)
4. Eljil, K.S., Qadah, G., Pasquier, M.: Predicting hypoglycemia in diabetic patients using time - sensitive artificial neural network. Int. J. Healthcare Inf. Syst. Inform. **11**(4), 70–88 (2016)
5. Favero, S.: First use of model predictive control in outpatient wearable artificial pancreas. Diabetes Care **37**, 1212–1215 (2014)
6. Finan, D.A., Palerm, C.C., Doyle, F.J., Seborg, D.E.: Effect of input excitation on the quality of empirical dynamic models for type 1 diabetes. Am. Inst. Chem. Eng. **55**(5), 1135–1146 (2009)
7. Gani, A., Gribok, A.V., Rajaraman, S., Ward, W.K., Reifman, J.: Predicting subcutaneous glucose concentration in humans: data-driven glucose modeling. IEEE Trans. Biomed. Eng. **56**(2), 246–254 (2009)

8. Georga, E., Protopappas, V., Ardigò, D., Marina, M., Zavaroni, I., Polyzos, D., Fotiadis, D.I.: Multivariate prediction of subcutaneous glucose concentration in type 1 diabetes patients based on support vector regression. IEEE J. Biomed. Health Inform. **17**(1), 71–81 (2013)
9. Georga, E., Protopappas, V., Fotiadis, D.I.: Glucose prediction in type 1 and type 2 diabetic patients using data driven techniques. In: INTECH (2011)
10. Hovorka, R., Allen, J.M., Elleri, D., Chassin, L.J., Harris, J., Xing, D., Kollman, C., Hovorka, T., Larsen, A.M.F., Nodale, M., De Palma, A.: Manual closed-loop insulin delivery in children and adolescents with type 1 diabetes: a phase 2 randomised crossover trial. Lancet **375**, 743–751 (2010)
11. Lee, H., Bequette, B.W.: A closed-loop artificial pancreas based on model predictive control: Human-family identification and automatic meal disturbance rejection. Biomed. Sig. Process. Control **4**, 347–354 (2009)
12. Magni, L., Raimondo, D.M., Man, C.D., Nicolao, G.D., Kovatchev, B., Cobelli, C.: Model predictive control of glucose concentration in subjects with type 1 diabetes: an in silico trial. In: IFAC (2008)
13. Markakis, M.G., Mitsis, G.D., Papavassilopoulos, G.P., Marmarelis, V.Z.: Model predictive control of blood glucose in type 1 diabetes: the principal dynamic modes approach. In: International IEEE EMBS Conference (2008)
14. Mougiakakou, S., Prountzou, A., Iliopoulou, D., Nikita, K.S., Vazeou, A., Bartsocas, C.S.: Neural network based glucose - insulin metabolism models for children with type 1 diabetes. In: Proceedings of the 28th IEEE, EMBS Annual International Conference New York City, USA, 30 August–3 September 2006
15. Oruklu, M.E., Cinar, A., Rollins, D.K., Quinn, L.: Adaptive system identification for estimating future glucose concentrations and hypoglycemia alarms. Automatica **48**(8), 1892–1897 (2012). Elsevier
16. Pappada, S.M., Cameron, B., Rosman, P.M.: Development of a neural network for prediction of glucose concentration in type 1 diabetes patients. J. Diabetes Sci. Technol. **2**(5), 792–801 (2008)
17. Quchani, S.A., Tahami, E.: Comparison of MLP and Elman neural network for blood glucose level prediction in type 1 diabetics. In: Ibrahim, F., Osman, N.A.A., Usman, J., Kadri, N.A. (eds.) 3rd Kuala Lumpur International Conference on Biomedical Engineering 2006. IFMBE, vol. 15. Springer, Heidelberg (2007)
18. Seron, M., Medioli, A., Goodwin, G.: A methodology for the comparison of traditional mpc and stohastic mpc in the context of the regulation of blood glucose levels in type 1 diabetes. In: Australian Control Conference (2016)
19. Soru, P., De Nicolao, G., Toffanin, C., Dalla Man, C., Cobelli, C., Magni, L., AP@ home consortium.: MPC based artificial pancreas: strategies for individualization and meal compensation. Ann. Rev. Control, **36**(1), 118–128 (2012)
20. Sparacino, G., Zanderigo, F., Corazza, S., Maran, A., Facchinetti, A., Cobelli, C.: Glucose concentration can be predicted ahead in time from continuous glucose monitoring sensor time-series. IEEE Trans. Biomed. Eng. **54**(5), 931–937 (2007)
21. Ståhl, F., Johansson, R.: Diabetes mellitus modeling and short-term prediction based on blood glucose measurements. Math. Biosci. **217**(2), 101–117 (2009)
22. Valletta, J.J., Chipperfield, A.J., Byrne, C.D.: Gaussian process modelling of blood glucose response to free-living physical activity data in people with type 1 diabetes. In: 31st Annual International Conference of the IEEE EMBS (2009)
23. Zainuddin, Z., Pauline, O., Ardil, C.: A neural network apporach in predicting blood glucose level for diabetic patients. Inte. J. Comput. Electr. Autom. Control Inf. Eng. **2**(3) (2009)

24. Zarkogianni, K., Mitsis, K., Litsa, E., Arredondo, M., Fico, G., Fioravanti, A., Nikita, K.: Comparative assessment of glucose prediction models for patients with type 1 diabetes mellitus applying sensors for glucose and physical activity. Int. Fed. Med. Biol. Eng. **53**(12), 1333–1343 (2015)
25. Zhao, C., Dassau, E., Jovanovic, L., Zisser, H.C., Doyle, F., Seborg, D.E.: Predicting subcutaneous glucose concentration using a latent-variable-based statistical method for type 1 diabetes mellitus. J. Diabetes Sci. Technol. **6**(3), 617–633 (2012)

Audit Trails in OpenSLEX: Paving the Road for Process Mining in Healthcare

Eduardo González López de Murillas[1]([✉]), Emmanuel Helm[2],
Hajo A. Reijers[1,3], and Josef Küng[4]

[1] Department of Mathematics and Computer Science,
Eindhoven University of Technology, Eindhoven, The Netherlands
{e.gonzalez,h.a.reijers}@tue.nl
[2] Research Department of e-Health, Integrated Care,
University of Applied Sciences Upper Austria, Hagenberg, Austria
emmanuel.helm@fh-hagenberg.at
[3] Department of Computer Science,
Vrije Universiteit Amsterdam, Amsterdam, The Netherlands
[4] Institute for Applied Knowledge Processing,
Johannes Kepler University, Linz, Austria
jkueng@faw.jku.at

Abstract. The analysis of organizational and medical treatment processes is crucial for the future development of the healthcare domain. Recent approaches to enable process mining on healthcare data make use of the hospital information systems' Audit Trails. In this work, methods are proposed to integrate Audit Trail data into the generic OpenSLEX meta model to allow for an analysis of healthcare data from different perspectives (e.g. patients, doctors, resources). Instead of flattening the event data in a single log file the proposed methodology preserves as much information as possible in the first stages of data extraction and preparation. By building on established standardized data and message specifications for auditing in healthcare, we increase the range of analysis opportunities in the healthcare domain.

Keywords: Process mining · Healthcare · Audit trails · Meta model

1 Motivation

Process mining provides an a-posteriori empirical method to discover process models from observed system behaviour. By applying the techniques of process mining to the healthcare domain, valuable insights regarding e.g. clinical practice, performance bottlenecks, and guideline compliance can be gained [13]. However, the healthcare domain presents certain challenges to traditional process mining approaches. Healthcare processes are highly dynamic, highly complex, increasingly multi-disciplinary, and, generally, ad hoc [12].

In [6] the authors tried to overcome the problems of distributed, heterogeneous data sources in healthcare by analyzing their minimum common

© Springer International Publishing AG 2017
M. Bursa et al. (Eds.): ITBAM 2017, LNCS 10443, pp. 82–91, 2017.
DOI: 10.1007/978-3-319-64265-9_7

ground – the Audit Trail (AT). Based on internationally agreed upon standards, these ATs contain only basic information about what happened even though these can be enriched to be useful for various business intelligence (BI) analysis.

This work will go further by solving a number of remaining issues of the previous approach (1) by explaining how process mining can be incorporated in the analysis of ATs, and (2) by providing the basis for automated process mining in different contexts via the OpenSLEX meta model [11].

The remainder of this paper is structured as follows: Sect. 2 introduces some background information. Then, Sect. 3 discusses the problem that motivates this work. The proposed approach is described in Sect. 4. Finally, Sect. 5 concludes the paper and gives a short outlook on future work.

2   Background

As has been mentioned before, the goal of this work is (1) to solve the data issues of the previous approach, and (2) to enable automated process mining on the extracted and transformed data. Before we attempt to tackle these challenges, it is necessary to introduce some background information about the data that we want to analyze, the ATs, and the meta model that we propose to represent the extracted information, OpenSLEX.

2.1   Standardized Audit Trails

The international non-profit organization *Integrating the Healthcare Enterprise* (IHE) aims to improve the integration and interoperability of healthcare IT systems. Founded in 1998 by radiologists and IT experts, it defines how established standards, like DICOM and HL7, can be implemented to overcome common interoperability problems in healthcare. The initial focus was on radiology but nowadays IHE covers use cases in different healthcare domains. Their *integration profiles* are the basis for systems of major vendors and national and international healthcare programs.

One of the basic IHE Integration Profiles dealing with IT infrastructure in healthcare, the *Audit Trail and Node Authentication* (ATNA) profile, defines how to build up a secure domain that provides patient information confidentiality, data integrity, and user accountability [7]. A secure domain can scale from department to enterprise to cross-enterprise size. To ensure user accountability, ATNA specifies the use of a centralized ARR where all Audit Messages are stored. In a joint effort IHE, HL7, DICOM, ASTM E31, and the Joint NEMA/COCIR/JIRA Security and Privacy Committee defined the structure of the Audit Messages using XML schema. The normative specification of the messages is defined in the DICOM standard PS3.15: A.5 Audit Trail Message Format Profile [2]. The original intention of ATNA event audit logging was to provide surveillance logging functions to detect security events and deviations from normal operations. It was not designed for forensic or workflow performance analysis. However, the integration profile states that forensic or workflow analysis logs may also use the same XML schema and IHE transactions [7] and recent developments propose the use of ATNA for keeping track of the whole workflow [3].

Cruz-Correia et al. analyzed the quality of hospital information systems (HISs) Audit Trails in Portugal [4]. They pointed out the potential use of ATs for the improvement of the HIS by applying data mining, machine learning, and process mining techniques. The authors highly recommend the use of standards to record ATs – including ATNA – but criticize the lack of completeness and overall data quality in the data they analyzed.

2.2 OpenSLEX Meta Model

Data extraction and transformation are, very often, the most time-consuming stages of a process mining project. The difficulty to tackle these tasks comes from the variability on data representations in which the original data can be found. Most of the applications of process mining in real-life systems provide ad-hoc solutions to the specific environment of application. Some examples of these systems are SAP [8,9,15] and other ERPs [10]. Nevertheless, efforts have been made to develop standards for data representation in process mining. The IEEE XES standard [1] is the most important example, being extensively used both in academic and industrial solutions. However, despite its success at capturing event data in an exchangeable format, something that this standard misses is the data perspective on the original system.

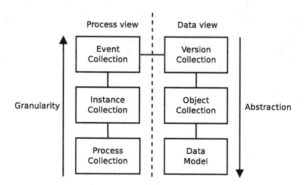

Fig. 1. Diagram of the OpenSLEX meta model at a high level.

With the purpose of mitigating the limitations of current event data representation standards, in previous work we proposed OpenSLEX [11], which provides a
meta model that takes into account not only the process view (*events, instances,*
and *processes*), but also the data view (*data models, objects,* and *object versions*). Figure 1 shows a high level description of the meta model, where we see
how granularity of data increases inversely proportional to the level of abstrac-
tion. In other words, a *data model* is a more abstract representation of the data

than the *objects* or the *object versions*, while the latter has a greater level of granularity than the *data model*. The same can be said about the process view, where *processes* are abstract descriptions of the *events*, which are much more granular data. A more detailed description of the meta model is available online[1].

Additionally, the fact that in this meta model the process side is combined with the data side allows to capture a richer snapshot of the system under study. Unlike other meta models, like the one proposed in XES, which requires the existence of a case notion to group events into process instances, OpenSLEX enables the adoption of different perspectives. Events are stored independently of any case notion. Afterwards, one or many case notions can be defined, generating the cases that will group events in different event logs. This is the key to enable multi-perspective process mining on the extracted data. The fact that not a single case notion is enforced during the data extraction phase avoids data loss. Additionally, it enables the application of automated techniques that correlate events in multiple ways to show different processes or perspectives coexisting in the same system.

To summarize, OpenSLEX provides a layer of standardization of the representation of data, while considering both process and data views, unlike other existing flatter event models. This makes it possible to decouple the application of analysis techniques from data extraction and transformation stages. Additionally, it enables smarter ways to analyze the information, considering the data side to enrich the process perspective.

3   Problem

The analysis of healthcare information systems is a challenging task. This is due to the unstructured nature of the processes, the large variety of data schemas and systems used in HISs, privacy issues, etc. [12]. These aspects make the application of process mining a time-consuming process. Some efforts have been made in order to standardize the way HISs communicate and transmit event data. The ATNA integration profile by IHE specifies the structure and content of messages needed for auditing. The profile describes how user accountability, especially regarding the use of protected health information (PHI), can be ensured by using a centralized Audit Record Repository (ARR) [7]. Therefore, every access to and transfer of PHI is recorded. Although the approach presented in [6] showed that the information recorded in audit trails is sufficient to enable process mining, several major issues remained unsolved. (1) The approach was not able to automatically identify traces. (2) The manually chosen, fixed trace identifier (the patient ID) lead to snapshot-problems and limited the possible process mining perspectives. (3) The hard-wired mapping of fields from ATNA audit messages to event logs sometimes lead to incorrect mappings.

In addition to tackling the mentioned issues this paper also addresses the topic of BI applications on top of ARRs, specifically in an automated manner. According to the integration profile the ATNA ARR is expected to have analysis

---

[1] https://github.com/edugonza/OpenSLEX/blob/master/doc/meta-model.png.

and reporting capabilities, but those capabilities are not defined as part of the profile [7]. The question is, *how* can BI applications be put on top of an ARR? And how can this be done in an automated way, assuming the data is stored and managed to be analyzed?

The following section describes how ATNA messages can be mapped into the OpenSLEX meta model, in order to tackle the issues of the previous approach, and the type of data transformation required to enable automated process mining.

4   Approach

The use of ATNA messages in order to extract event data for process mining has been previously demonstrated in [6]. Now we aim at stepping up in the level of generalization, using the OpenSLEX meta model as an intermediate format for data collection. This meta model can be seen as a data schema for a data warehouse, acting as an ARR, capturing event data, together with data objects, data models, and historical data, ready to be exploited by the existing process mining techniques.

4.1   Mapping of ATNA Messages to OpenSLEX

The specific characteristics of ATNA messages makes them great candidates for event data extraction. Figure 2 shows how different fields of the ATNA message can be mapped to fields of the OpenSLEX meta model. The minimally required attributes in order to obtain events are activity names and timestamps. These two attributes can be directly mapped to the ATNA message's fields *EventID* and *EventDateTime*, respectively. In addition, *Active Participant* fields such as *UserID* and *UserName* show valuable resource data to enrich the events. However, what makes ATNA messages specially attractive from the process mining perspective is the presence of *Participant Object* data. The fields within this part of the message contain not only object data information such as role (*ParticipantObjectTypeCodeRole*) and life-cycle (*ParticipantObjectDataLifeCycle*), but also object type (*ParticipantObjectTypeCode*) and unique object identifiers (*ParticipantObjectID*), which enable the traceability of data changes and behavior at object level. Additionally, detailed value pair data (*ParticipantObjectDetail*) of the participating object can be present. Such key-value pairs represent a snapshot of the relevant attributes of a participant object at the time of occurrence of the event, which can be seen as an object version. Object versions reveal the evolution of objects through time, and are related to the events that caused the modifications.

Data extraction and transformation are difficult tasks that require a significant amount of domain knowledge to be carried out. It is common that, during this transformation of data, choices are made that affect the final picture we obtain of the system under study. Considering the ATNA message fields we just discussed, we seem to be able to capture event information, which may be

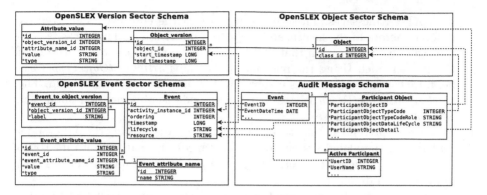

Fig. 2. The dashed lines show the mapping of the fields of Audit Messages to the OpenSLEX meta model.

mapped into the OpenSLEX corresponding elements. The next section explains the transformation of the captured event data in order to infer new information. This will allow to obtain a more complete picture of the whole system.

4.2 Transformation of ATNA Messages

OpenSLEX provides the meta model to shape the extracted information, with the purpose of minimizing the impact of the data extraction and transformation stages on the result of the analysis. Transforming the ATNA messages into this new representation enables the application of process mining without any semantic loss of the original data. This is achieved by considering the data view in addition to the process view, avoiding flattening multidimensional data into simple event logs. Figure 3 shows the steps in the data transformation process in order to capture a picture as close as possible to the original system:

(a) Figure 3a shows the situation in which we only obtained event information from the system. This matches the situation we face when dealing solely

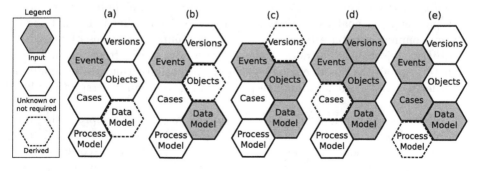

Fig. 3. Inference of the missing elements in the meta model, starting from the events (a) and finishing mining a model (e).

with ATNA messages. These messages are, in the end, events emitted by different actors in the healthcare ecosystem under study. These events contain valuable information that will let us infer some of the other sectors of the meta model. Using the information in the *ParticipantObjectTypeCode* field, we can infer the classes involved in the data model of the system. The *ParticipantObjectDetail* key-value pairs provide the information about the attributes of such classes.

(b) Figure 3b represents the next step in which, after discovering the data model underneath the global system, object instances are mapped into it. To do so, the field *ParticipantObjectID* helps us to identify the unique entities that exist for each data class.

(c) Figure 3c depicts the subsequent step, in which we infer the object versions involved in the process. These object versions are object snapshots at a certain moment in time, and they can be reconstructed by matching the key-value pairs in the *ParticipantObjectDetail* field of the ATNA events, to the object id obtained in the previous step (Fig. 3c). Reconstructing these event versions will help us understand the object's life-cycle within the process. And not only that, but applying primary and foreign key discovery techniques [14,16] will make possible to uncover links between objects belonging to different classes. In the next step we will see how these connections can be exploited to correlate events in different ways.

(d) So far we have been able to capture events, infer the data model, and extract object and object version information from the data. However, a case notion is needed in order to group events in cases. These cases will build the event logs necessary for process the mining analysis. It is in this step (Fig. 3d) in which one of the main benefits of this technique arises: the independence of the case notion from the event capturing phase. Events can be correlated in many different ways. One of them is to select a common attribute as a case identifier, which is the most common way to build logs nowadays. However, our meta model gives us an advantage with respect to traditional methods: the existence of links between events and object versions. As has been described in the previous step, relations between object versions can be discovered. This means that objects can point to others, as they do in databases with foreign keys, e.g., a treatment object points to a doctor and a patient. This enables a new way to correlate events that might not share any common attribute (doctor and patient events), by means of a linking object (treatment). The data model structure discovered from the data will determine all the possible combinations (case notions) that can be made in order to build event logs, making possible to have a multi-perspective view on the data.

(e) Only when case notions have been discovered, and the logs representing the different views have been built, we can proceed with the process discovery. Figure 3e shows the step in which process models are discovered using existing process mining techniques. This is the step that enables further analysis of the data combining process and data views in a centralized and comprehensive manner.

The steps above describe the transformation of ATNA messages into the OpenSLEX format. Sometimes this requires to infer the missing bits of information from the ones available in the ATNA message's attributes. As a result, we obtain a global view of the data and process sides, minimizing data loss during the extraction and transformation processes. The following section discusses in further detail the benefits of this transformation.

5   Conclusion and Future Work

The aim of the presented work is twofold. (1) To solve the issues found in the previous approach, trying to decouple the data extraction of audit trails from the application of analysis techniques. To do so, a mechanism is provided to integrate audit trails into the OpenSLEX meta model proposed in [11]. (2) To enable the application of new process mining techniques in an automated way, without the need for extensive domain knowledge, thanks to the standardization of the data representation proposed by OpenSLEX.

In order to solve the issues found in the previous approach, a new way to extract data from ATNA messages had to be developed. This new approach integrates ATNA messages into the OpenSLEX meta model to obtain a global, multi-perspective view of the data, reducing data loss as much as possible. This is achieved by a comprehensive data extraction process. Additionally, inference of incomplete data becomes possible by analyzing the participant objects indicated within the ATNA messages. All these data aspects of the process are rarely exploited in an integrated way by the existing event models such as XES. The proposed OpenSLEX meta model combines these data aspects with the process view to provide an integrated and complete picture of the system under study. On top of that, it provides the tools to discover the underlying data schema of the fragmented pieces of data captured in an heterogeneous environment such as current healthcare information systems.

Another one of the benefits of this transformation is that, collecting all the information in a standardized meta model enables the application of analysis techniques independently of the origin of data. Decoupling analysis from data extraction makes it possible to apply many analysis techniques with minimum effort. The standardization of queries, the application of automated machine learning, automated trace identification, and multi-perspective analysis are some of the techniques that become possible to apply when using a standardized and complete representation of data like the one that OpenSLEX provides.

The presented work is a first step in order to enhance the possibilities of process mining in healthcare. The fact that now we are able to process ATNA messages, allows us to exploit new possibilities, like the recent Standardized Operational Log of Events (SOLE). SOLE is a technical framework supplement of IHE that proposes the use of the ATNA message specification and IHE transaction (ITI-20) to communicate events regarding the whole radiological workflow [3]. SOLE requires the use of the SWIM (SIIM Workflow Initiative in Medicine) workflow lexicon to define the semantics of the audit messages. This term list was created by the Society for Imaging Informatics (SIIM) to "improve the quality of information available to people managing imaging departments" [5] and was recently added to the RadLex[2] ontology. With audit messages describing events from start to end, the entries in an ARR can be used to analyze the complete workflow of an imaging facility. We plan to further develop our approach and do a case study based on real life event data from a radiological facility using the SOLE profile.

References

1. IEEE Standard for eXtensible Event Stream (XES) for Achieving Interoperability in Event Logs and Event Streams. IEEE Std 1849–2016, pp. 1–50, November 2016
2. National Energy Marketers Association: Nema PS3/ISO 12052, digital imaging and communications in medicine (dicom) standard. Website (2016). http://medical.nema.org/
3. Infrastructure Technical Committee: Standardized operational log of events (sole). IHE Radiology Technical Framework Supplement Rev. 1.0 Draft for Public Comment, 51 (2017)
4. Cruz-Correia, R., Boldt, I., Lapão, L., Santos-Pereira, C., Rodrigues, P.P., Ferreira, A.M., Freitas, A.: Analysis of the quality of hospital information systems audit trails. BMC Med. Inf. Decis. Making 13(1), 84 (2013)
5. Erickson, B.J., Meenan, C., Langer, S.: Standards for business analytics and departmental workflow. J. Digital Imaging 26(1), 53–57 (2013)
6. Helm, E., Paster, F.: First steps towards process mining in distributed health information systems. Int. J. Electron. Telecommun. 61(2), 137–142 (2015)
7. IHE: Audit trail and node authentication (atna). IHE IT Infrastructure (ITI) Technical Framework, vol. 1, Integration Profiles ITI-TF-1, 71–83 (2016)
8. Ingvaldsen, J.E., Gulla, J.A.: Preprocessing support for large scale process mining of SAP transactions. In: Hofstede, A., Benatallah, B., Paik, H.-Y. (eds.) BPM 2007. LNCS, vol. 4928, pp. 30–41. Springer, Heidelberg (2008). doi:10.1007/978-3-540-78238-4_5
9. ER, M., Astuti, H.M., Wardhani, I.R.K.: Material movement analysis for warehouse business process improvement with process mining: a case study. In: Bae, J., Suriadi, S., Wen, L. (eds.) AP-BPM 2015. LNBIP, vol. 219, pp. 115–127. Springer, Cham (2015). doi:10.1007/978-3-319-19509-4_9
10. Mueller-Wickop, N., Schultz, M.: ERP event log preprocessing: timestamps vs. accounting logic. In: Brocke, J., Hekkala, R., Ram, S., Rossi, M. (eds.) DESRIST 2013. LNCS, vol. 7939, pp. 105–119. Springer, Heidelberg (2013). doi:10.1007/978-3-642-38827-9_8
11. González López de Murillas, E., Reijers, H.A., Aalst, W.M.P.: Connecting databases with process mining: a meta model and toolset. In: Schmidt, R., Guédria, W., Bider, I., Guerreiro, S. (eds.) BPMDS/EMMSAD -2016. LNBIP, vol. 248, pp. 231–249. Springer, Cham (2016). doi:10.1007/978-3-319-39429-9_15
12. Rebuge, Á., Ferreira, D.R.: Business process analysis in healthcare environments: a methodology based on process mining. Inf. Syst. 37(2), 99–116 (2012)
13. Rojas, E., Munoz-Gama, J., Sepúlveda, M., Capurro, D.: Process mining in healthcare: a literature review. J. Biomed. Inf. 61, 224–236 (2016)

---

[2] RadLex ontology: http://www.radlex.org.

14. Sismanis, Y., Brown, P., Haas, P.J., Reinwald, B.: Gordian: efficient and scalable discovery of composite keys. In: Proceedings of the 32nd International Conference on Very Large Data Bases, pp. 691–702. VLDB Endowment (2006)
15. Štolfa, J., Kopka, M., Štolfa, S., Koběrský, O., Snášel, V.: An application of process mining to invoice verification process in SAP. In: Abraham, A., Krömer, P., Snášel, V. (eds.) Innovations in Bio-inspired Computing and Applications. AISC, vol. 237, pp. 61–74. Springer, Cham (2014)
16. Zhang, M., Hadjieleftheriou, M., Ooi, B.C., Procopiuc, C.M., Srivastava, D.: On multi-column foreign key discovery. Proc. VLDB Endowment **3**(1–2), 805–814 (2010)

Riemannian Geometry
in Sleep Stage Classification

Elizaveta Saifutdinova[1,2(✉)], Václav Gerla[3], and Lenka Lhotská[3,4]

[1] Faculty of Electrical Engineering, Department of Cybernetics,
Czech Technical University in Prague, Prague, Czech Republic
saifueli@fel.cvut.cz
[2] National Institute of Mental Health, Prague, Czech Republic
[3] Czech Institute of Informatics, Robotics and Cybernetics,
Czech Technical University in Prague, Prague, Czech Republic
[4] Faculty of Biomedical Engineering, Czech Technical University in Prague,
Prague, Czech Republic

Abstract. The study is devoted to the sleep stage identification problem. Proposed method is based on calculation of covariance matrices from segments of multi-modal recordings. Mathematical properties of the extracted covariance matrices allow to define a distance between two segments - a distance in a Riemannian manifold. In the paper we tested minimum distance to a class center and k-nearest-neighbours classifiers with the Riemannian metric as a distance between two objects, and classification in a tangent space to a Riemannian manifold. Methods were tested on the data of patients suffering from sleep disorders. The maximum obtained accuracy for KNN is 0.94, for minimum distance to a class center it is only 0.816 and for classification in a tangent space is 0.941.

Keywords: Sleep EEG · Sleep stages · Riemannian geometry · Classification

1 Introduction

The study addresses the problem of automated sleep stage identification. It is in the area of interest of many researchers for several decades. Polysomnography (PSG) recorded during a night helps clinicians to analyze sleep process. This diagnostic test monitors many body functions including brain activity (EEG), eye movements (EOG), muscle activity (EMG), heart rhythm (ECG), and breathing function. With the development of measurement techniques and medical recording devices, it is possible to collect more data during sleep. It plays a crucial role in sleep disorders investigation. The general standard of sleep PSG evaluation developed by American Academy of Sleep Medicine (AASM) [5] defines four basic sleep stages, which repeatedly change one another. Rapid eye movement (REM) sleep is characterized by high eye activity and muscle atonia. All other stages are called non-rapid eye movement (NREM) and could be

© Springer International Publishing AG 2017
M. Bursa et al. (Eds.): ITBAM 2017, LNCS 10443, pp. 92–99, 2017.
DOI: 10.1007/978-3-319-64265-9_8

divided into light and deep sleep. Deep sleep stage S3 is characterized by prevailing of slow waves. Light sleep stages S1 and S2 are commonly specified as EEG activity in alpha and theta frequency bands. S2 sleep stage is hall-marked by special EEG patterns like sleep spindles and K-complexes. Multimodal PSG recording is utilized for a sleep stage identification. Sleep stage investigation is critical in the examination of a patient with sleep disorders. It can give a lot of details about the severity of syndromes, predict the development of the disorder or explain causes of syndromes [1, 6, 9].

During previous years, there were presented a lot of studies devoted to a sleep stage classification. Some of them use support vector machine, Hidden Markov model or neural networks as a classification engine [12, 13, 17]. However, not only EEG are used to detect stages of sleep process. Thus only cardio and breathing activity helps to identify REM and NREM sleep [3]. Analyzing EOG channels serves for REM stage detection [5]. Most of the studies calculate a feature set describing the data segment. For example, there were utilized spectral features, coefficients of wavelet transformation or statistical characteristics as features for classification [4, 8, 11, 12]. Usually, they are extracted from separate channels or their combination. Growing number of recorded channels increases the number of features. Widespread methods of optimal feature selection are based on statistic, for example, fast correlation based feature selection, t-test and Fisher score [15].

Using a spatial information is not so common approach in sleep stage classification though it could bring additional information about mutual dependency between channels. In this case, a data segment is represented by a covariance matrix. It could be applied to the random number of recorded channels. Estimate of covariance matrix has interesting mathematical properties. It is a symmetric and positive-definite matrix, it belongs to a Riemannian manifold [16]. A distance defined in a Riemannian space gives a measure of similarity between two SPD matrices. Another interesting property is that a tangent space to a manifold is Euclidean. That allows using state-of-the-art classification methods to object projections. Previously spatial features were used for brain-computer interface classification tasks [2, 7]. There were compared EEG short segments recorded using 22 electrodes. Lately, that approach was implemented for sleep stage classification by Li and Wong [10]. Instead of covariance between channels, authors used power spectral covariance matrices [10]. In the study, they used four referenced channels. A classification was performed using the k-nearest-neighbour (KNN) method, the distance between data segments was a combination of distances between covariances at different frequencies.

In the paper, we are going to work with covariance matrices of the recorded channels without using spectral information to identify sleep stages. Data obtained from different sources will be used for classification. A measure of similarity between two segments is just a distance between covariance matrices of these segments. The aim of the paper is to test sleep stage identification methods based on classification of covariance matrices obtained from the data.

2   Materials and Methods

2.1   Covariance Matrix

Covariance matrix shows dependence strength between channels. Each element at position (i, j) is the covariance between channels i and j. In comparison with correlation, it is measured in the units of the original variables. For a signal $X \in R^{n \times m}$ of n channels and length m estimate of covariance matrix C is calculated by Sample Covariance Matrix (SCM) expressed by the Eq. 1.

$$C = \frac{1}{n-1} X X^T. \tag{1}$$

This method is unbiased, and it assumes that the number of samples m is much greater than the number of channels n. Obtained estimation of covariance matrix $C \in R^{n \times n}$ is symmetric and positive definite. A traditional way to measure a distance between two matrices A and $B \in R^{n \times n}$ is to calculate a Frobenius norm, which could be expressed in the equation:

$$\|A, B\|_F = (\sum_{i=1}^{n} \sum_{i=1}^{n} \|a_{ij} - b_{ij}\|^2)^{\frac{1}{2}} = \sqrt{tr[(A - B)(A - B)^H]}. \tag{2}$$

2.2   Riemannian Geometry

Since the matrix C is symmetric and positive definite (SPD), we can define a differentiable Riemannian manifold of covariance matrices [10]. That means that there is defined an inner product on each tangent space to the manifold which also varies smoothly from point to point. Then the distance between two points s_i and s_j on the manifold is an inner product on the tangent space of the manifold at a point P:

$$\langle s_i, s_j \rangle = tr(P^{-1} s_i P^{-1} s_j). \tag{3}$$

Defining a distance between the SPD matrices allows specifying a mean value of the SPD matrices set. There are several approaches to do that. The simplest approach is to calculate the arithmetic mean M of the matrices C_i using Eq. 4.

$$M = \frac{1}{K} \sum_{i=1}^{K} C_i. \tag{4}$$

However, due to properties of a Riemannian space, estimation of mean value obtained with Eq. 4 has poor properties and not reliable in practical applications. For getting a representative of the class, the most useful approach is to calculate a geometric mean. In Riemannian geometry, it minimizes the sum of squared hyperbolic distances to all elements from itself. The geometric mean M of K SPD matrices $C = \{C_1, ..., C_K\}$ is calculated iteratively starting with a smart guess. For instance, it could be initialized with arithmetic mean. On each iteration it is updated as follows [14]

$$M \leftarrow M^{\frac{1}{2}} \exp[\frac{1}{k}\sum_k \ln(M^{-\frac{1}{2}}C_k M^{-\frac{1}{2}})]M^{-\frac{1}{2}}, \qquad (5)$$

until converges

$$\| \sum_k \ln(M^{-\frac{1}{2}}C_k M^{-\frac{1}{2}})]M^{-\frac{1}{2}}\|_F < \varepsilon. \qquad (6)$$

This algorithm converges linearly. It may not converge in a case of high dimensional matrices very distant from each other. Riemann distances to geometric mean do not have a symmetric distribution.

2.3   Classification in Riemannian Geometry

We are going to use simple and efficient distance-based classifiers to identify sleep stages. Covariance matrices are extracted from defined segments of sleep PSG data by the equation described in Subsect. 2.1. For obtained covariance matrices there is defined distance in a Riemannian manifold. K-nearest-neighbours classifier could be simply applied to the data [10]. A label of a new element is arranged by the label of the majority among K closest elements, where the number of neighbors K is determined. It is robust to a noise and sufficient for a large dataset. Another multi-class classifier uses a minimum distance to a class center [2]. Defining representative of a class we calculate a geometric mean by the method described in Subsect. 2.2 as a class center. A new object is labeled with a label of the closest class center. A tangent space to a Riemannian manifold is a Euclidean space [2]. A matrix object is represented as a vector and state-of-the-art classifiers like logistic regression.

3   Experimental Results

3.1   Data Description

Data of seven patients with sleep disorders were recorded in National Institute of Mental Health, Czech Republic. There were recorded 19 EEG channels with electrodes placed by 10–20 standard (Fp1, Fp2, F3, F4, C3, C4, P3, P4, O1, O2, F7, F8, T3, T4, T5, T6, Fz, Cz, Pz, M1, M2.), breathing channel, ECG, EOM and chin EMG channels. Artifact-free segments were manually extracted from the data. The data were divided into segments of length 10 s.

3.2   Results

Classification methods were investigated on the real PSG recordings. Patients data were used separately for analysis. Classification methods were tested by cross-validation scheme with ten folds; each fold contains the same proportion of labels as the whole recording. The procedure was repeated for each fold as a

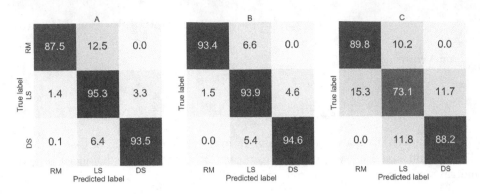

Fig. 1. Confusion matrix for a classification of REM (RM), S1 and S2 (LS) and S3 (DS) stages using all EEG and all non-EEG channels. Matrices correspond to classification in a tangent space (A) and using k-nearest-neighbor (B) and minimum distance to a class center (C) classifiers with distance in Riemannian space.

validation set and all the remaining folds as a training data. Accuracy value was calculated on each iteration such as

$$Acc = \frac{TC}{N},\tag{7}$$

where TC is a number of correctly classified elements and N is the total number of elements. Accuracy is in the range $[0, 1]$ and 1 corresponds to the best performance. Confusion matrices were used for results evaluation as well. An element at position (i, j) is equal to the number of observations known to be in class i but predicted to be in class j.

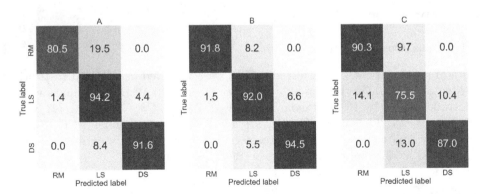

Fig. 2. Confusion matrix for a classification of REM (RM), S1 and S2 (LS) and S3 (DS) stages using selected EEG and all non-EEG channels. Matrices correspond to classification in a tangent space (A) and using k-nearest-neighbor (B) and minimum distance to a class center (C) classifiers with distance in Riemannian space.

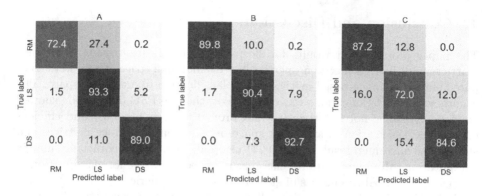

Fig. 3. Confusion matrix for a classification of REM (RM), S1 and S2 (LS) and S3 (DS) stages using selected EEG and EOG channels. Matrices correspond to classification in a tangent space (A) and using k-nearest-neighbor (B) and minimum distance to a class center (C) classifiers with distance in Riemannian space.

Covariance matrices were extracted from 10-second artifact-free segments. Each segment contains EEG, EMG, EOG and breathing channels. Stages S1 and S2 were joined in due the small number of clean S1 segments. Classification results using KNN (k = 5), the minimum distance to a class center classifiers with Riemannian distance and classification in a tangent space are shown in the Fig. 1.

In practice, there is possible to define sleep stages by visual analysis using electrodes in frontal lobe and EOG channels. In presented method using a smaller number of channels simplifies the calculations. We reduced the number of EEG channels to C3, C4, Cz, F3, F4, Fp1, Fp2, Fz. Covariance matrices consisted of selected EEG and two EOG channels. Confusion matrices of classification using selected EEG and EOG channels is on the Fig. 2. Results are significantly lower than in the previous test. The average accuracy of classification using KNN is 0.91, for minimum distance classifier it is 0.774 and for classification in a tangent space, it is 0.899.

Adding all other non-EEG channels into covariance matrix improves the results. The average accuracy for KNN is 0.929, for minimum distance to a class center it is only 0.816 and for classification in a tangent space is 0.927. The Fig. 3 presents confusion matrices for that classification (Table 1).

Table 1. Summary of experiments and average accuracy of classification

	All channels	Selected EEG and EOG	Selected EEG and non EEG
KNN	0.94	0.91	0.927
Minimum distance	0.794	0.774	0.816
Classification in a tangent space	0.941	0.899	0.927

4   Conclusion and Discussion

The paper provides test results for sleep stage classification using covariance matrices and their geometric properties in a Riemannian space. There were compared simple distance-based methods and classification in a tangent space. The accuracy of the classifications was obtained on the real PSG data. KNN showed good performance not less than 0.9 and proved itself as a robust and efficient classifier. Classification in a tangent space depends on covariance matrices and showed the maximum result equals 0.941 using all channels for covariance matrix computation. Minimum distance classifier is not appropriate for classification of sleep stages, probably, clusters of sleep data are not spheric.

Using only covariance matrices allows classifying the sleep data with satisfactory accuracy. Developing new recording devices allow to register more data during sleep and makes detection more accurate without using spectral information. This method applies to a random number of channels and easy to implement. This approach could provide quick evaluation for a new recording based on the data recorded previously.

Future research could be devoted to the clustering problem. The complex shape of clusters assumes using advanced unsupervised methods. Probably, representation of a cluster as a composition of smaller subclusters could be a good solution. Another option is investigating the cluster structure in a tangent space. Another topic of the research is a detection of abnormal patterns in sleep night recordings.

Acknowledgment. Research of E. Saifutdinova was supported by the project No. SGS17/135/OHK4/2T/13 of the Czech Technical University in Prague. This work was also supported by the project "National Institute of Mental Health (NIMH-CZ)", grant number ED2.1.00/03.0078 and the European Regional Development Fund and by the project Nr. LO1611 with a financial support from the MEYS under the NPU I program. Research of V. Gerla and L. Lhotska was partially supported by the project "Temporal context in analysis of long-term non-stationary multidimensional signal", register number 17-20480S of the "Grant Agency of the Czech Republic."

References

1. Arfken, C., Joseph, A., Sandhu, G., Roehrs, T., Douglass, A., Boutros, N.: The status of sleep abnormalities as a diagnostic test for major depressive disorder. J. Affect. Disord. **156**, 36–45 (2014). http://www.sciencedirect.com/science/article/pii/S0165032713008434
2. Barachant, A., Bonnet, S., Congedo, M., Jutten, C.: Multiclass brain computer interface classification by riemannian geometry. IEEE Trans. Biomed. Eng. **59**(4), 920–928 (2012)
3. Fonseca, P., Long, X., Radha, M., Haakma, R., Aarts, R.M., Rolink, J.: Sleep stage classification with ecg and respiratory effort. Physiol. Meas. **36**(10), 2027 (2015)
4. Hassan, A.R., Bhuiyan, M.I.H.: Automatic sleep scoring using statistical features in the emd domain and ensemble methods. Biocybernetics Biomed. Eng. **36**(1), 248–255 (2016)

 5. Iber, C.: American Academy of Sleep Medicine: The AASM Manual for the Scoring of Sleep and Associated Events: Rules, Terminology and Technical Specifications. American Academy of Sleep Medicine (2007)
 6. Jones, S.G., Riedner, B.A., Smith, R.F., Ferrarelli, F., Tononi, G., Davidson, R.J., Benca, R.M.: Regional reductions in sleep electroencephalography power in obstructive sleep apnea: a high-density eeg study. Sleep **37**(2), 399 (2014)
 7. Kalunga, E.K., Chevallier, S., Barthlemy, Q., Djouani, K., Monacelli, E., Hamam, Y.: Online ssvep-based bci using Riemannian geometry. Neurocomputing **191**, 55–68 (2016)
 8. Khalighi, S., Sousa, T., Pires, G., Nunes, U.: Automatic sleep staging. a computer assisted approach for optimal combination of features and polysomnographic channels. Expert Syst. Appl. **40**(17), 7046–7059 (2013)
 9. Lafortune, M., Gagnon, J.F., Martin, N., Latreille, V., Dub, J., Bouchard, M., Bastien, C., Carrier, J.: Sleep spindles and rapid eye movement sleep as predictors of next morning cognitive performance in healthy middle-aged and older participants. J. Sleep Res. **23**(2), 159–167 (2014). http://dx.doi.org/10.1111/jsr.12108
10. Li, Y., Wong, K.M.: Riemannian distances for signal classification by power spectral density. IEEE J. Sel. Top. Sig. Process. **7**(4), 655–669 (2013)
11. Lotte, F., Congedo, M., Lecuyer, A., Lamarche, F., Arnaldi, B.: A review of classification algorithms for eeg-based brain computer interfaces. J. Neural Eng. **4**(2), R1 (2007). http://stacks.iop.org/1741-2552/4/i=2/a=R01
12. Motamedi-Fakhr, S., Moshrefi-Torbati, M., Hill, M., Hill, C.M., White, P.R.: Signal processing techniques applied to human sleep eeg signals a review. Biomed. Sign. Process. Control **10**, 21–33 (2014)
13. Peker, M.: A new approach for automatic sleep scoring. combining taguchi based complex-valued neural network and complex wavelet transform. Comput. Methods Programs Biomed. **129**, 203–216 (2016)
14. Pennec, X., Fillard, P., Ayache, N.: A riemannian framework for tensor computing. Technical report 5255 (2004)
15. Şen, B., Peker, M., Çavuşoğlu, A., Çelebi, F.V.: A comparative study on classification of sleep stage based on eeg signals using feature selection and classification algorithms. J. Med. Syst. **38**(3), 18 (2014)
16. Yger, F., Berar, M., Lotte, F.: Riemannian approaches in brain-computer interfaces: a review. IEEE Trans. Neural Syst. Rehabil. Eng. **PP**(99), 1 (2016)
17. Zhang, Y., Zhang, X., Liu, W., Luo, Y., Yu, E., Zou, K., Liu, X.: Automatic sleep staging using multi-dimensional feature extraction and multi-kernel fuzzy support vector machine. J. Healthc. Eng. **5**(4), 505–520 (2014)

The Use of Convolutional Neural Networks in Biomedical Data Processing

Miroslav Bursa$^{(\boxtimes)}$ and Lenka Lhotska

Czech Institute of Informatics, Robotics and Cybernetics,
Czech Technical University in Prague, Prague, Czech Republic
{miroslav.bursa,lenka.lhotska}@cvut.cz

Abstract. In this work, we study the use of convolutional neural networks for biomedical signal processing. Convolutional neural networks show promising results for classifying images when compared to traditional multilayer perceptron, as the latter do not take spatial structure of the data into an account.

Cardiotocography (CTG) is a monitoring of fetal heart rate (FHR) and uterine contractions (UC) used by obstetricians to assess fetal wellbeing. Because of the complexity of FHR dynamics, regulated by several neurological feedback loops, the visual inspection of FHR remains a difficult task. The application of most guidelines often result in significant inter-and intra-observer variability.

Convolutional neural network (CNN, or ConvNet) is inspired by the organization of the animal visual cortex.

In the paper we are applying continuous wavelet transform (CWT) to the UC and FHR signals with different levels of time/frequency detail parameter and in two different resolutions. The output 2D structures are fed to convolutional neural network (we are using Tensorflow framework [1]) and we are minimizing the *cross entropy* function.

On the testing dataset (with pH threshold at 7.15) we have achieved the accuracy of 94.1% which is a promising result that needs to be further studied.

Keywords: Data mining · Cardiotocography · Intrapartum · Signal processing · Convolutional neural networks

1 Introduction

In this work, we have studied the use of convolutional neural networks for biomedical signal processing. According to our best knowledge, this kind of networks has not been used with the cardiotocography data for asphyxia (fetal acidosis) prediction. Convolutional neural networks show promising results for classifying images when compared to traditional multilayer perceptron, as the latter do not take spatial structure of the data into an account and significantly suffer from the *curse of dimensionality*. One of the issues is that convolutional neural networks usually expect a kind of image on the input, but the signals are only

© Springer International Publishing AG 2017
M. Bursa et al. (Eds.): ITBAM 2017, LNCS 10443, pp. 100–119, 2017.
DOI: 10.1007/978-3-319-64265-9_9

1D structure. This can be resolved by several transformations. Two of them are described further.

Cardiotocography (CTG) is a monitoring of fetal heart rate (FHR) and uterine contractions (UC) used by obstetricians to assess fetal well-being during the process of delivery. Many attempts to introduce methods of automatic signal processing, evaluation and synthesis with clinical data have appeared during the last decades [3,9], however still no significant progress similar to that in the domain of adult heart rate variability is visible. Because of the complexity of FHR dynamics, regulated by several neurological feedback loops, the visual inspection of FHR still remains an uneasy task. The application of most clinical guidelines often result in significant inter- and intra-observer variability. The data used in this work originate from the freely available CTG-UHB database [4], available from physionet.org [7].

Convolutional neural network (CNN, or ConvNet) belong to feed-forward artificial neural network where the connectivity pattern between its neurons is inspired by the organization of the animal visual cortex. In the CNN, the neurons are formed into 3D volumes, connected to a small receptive feed. The network consists of several convolutional layers, connected to a fully connected (subsampling layer), so it can model complex non-linear relationships.

In the paper we are applying continuous wavelet transform (CWT) to the UC and FHR signals with different levels of time/frequency detail parameter and in two different resolutions. The output 2D structures are fed to convolutional neural network (we are using Tensorflow framework [1]) and we are using the *cross entropy* function as a criterion to be minimized during the learning process.

1.1   Motivation and Clinical View

Severe asphyxia (fetal acidosis) during the process of childbirth can lead to various complications, resulting into serious brain damage of the neonate as the brain is the most sensitive to oxygen deprivation. In case of fetal acidosis early detection, Cesarean section might be indicated in time in order to speed up the whole process and reduce the time the newborn spends in oxygen insufficiency. During the delivery, certain (but not severe) asphyxias are perfectly normal and natural part of the childbirth process.

On the other hand, the doctors aim to avoid unnecessary Cesarean sections as much as possible. In fact, they perform multiobjective optimization, while minimizing the overall risk. This is, however, hard to evaluate quantitatively and the doctors make their best decision based on their experience and guidelines.

The FHR is traditionally investigated using Fourier transform, ofen with predefined frequency bands powers and corresponding LF/HF ratio. Many modern trends and developments in the fields of signal processing and machine learning have been used, such as autocorrelation and spectral analysis, using time-scales and scale-free paradigms. Complex parameters are usually based on variations of entropy rates and multifractal and scattering analyses. The usual problem of such works is that the works use proprietary and/or small databases. In [6] the authors used fractal analysis and Hurst parameter for the HR variability.

In [11] the authors successfully used (sparse) support vector machines for FHR classification.

The detection and decision process is of course much more complicated and the prediction requires long-term praxis of the expert, and is usually based on the cardiotocography signal (CTG) that is being monitored during the delivery. Mostly, the evaluation is done visually, following guidelines maintained by international and national scientific societies.

Because of the complexity of FHR dynamics, regulated by several neurological feedback loops, the visual inspection of FHR remains a difficult task. The application of most guidelines often result in significant inter- and intra-observer variability[1]. The study on the expert obstetrician's agreement has been also published [5,8,12]. The high values of the aforementioned variability is the driving force of the search for an objective evaluation of the cardiotocography signal.

Many studies showed that cardiotocography signal itself is not sufficient and supplemental clinical information is needed for making the correct decision – such as APGAR score, biochemical markers (pH value, base excess, base deficit). Often family and mother anamneses, risk factors and other factors need to be added [3]. These information are often available in the hospital information system, almost never in a format that allows direct or easy processing. In this work this information is not presented to the convolutional network.

The cardiotocography signal processing is a crucial part of Decision Support System which fits the overall aim of this work: to ease the work of obstetricians and to help them improve their decision to better preserve the health of the mother and the neonate.

2   Data and Methodology

2.1   Input Dataset Description

The process of CTG signal selection (and restriction) is not described within this paper, as the rules are available together with the CTG signal database made freely available for comparative studies[2].

We are also omitting the detailed DB parameters, as they are available within the related publication [4]. The database contains 552 singleton recordings (selected from the total of 9164) with mean pH of 7.23. It consists of 46 Cesarean sections and 506 vaginal deliveries (44 operative).

In the literature you can see many parameters and different values (and rules) to divide the cases (recordings) into normal and pathological classes. In [4] you can find a comprehensive overviews of the thresholds and rules used.

---

[1] Moreover, the results often differ by location and/or country. You can find more information in the works of Spilka et al., i.e. [12].

[2] The open-access database [4] is freely available at the following link: http://www.physionet.org/physiobank/database/ctu-uhb-ctgdb/. We therefore encourage other researchers to take advantage of this free database available.

In this paper we have used the decision threshold as follows: the pathological records were considered those having the value of pH $\leq$ 7.15, the normal recordings were those having pH value > 7.15.

2.2 Preprocessing and Feature Extraction

In this part we describe the preprocessing of the input data and the subsequent feature extraction and subsampling.

Each recording consists of two signals, fetal heart rate (*FHR*) and uterine contraction (*UC*). The FHR has been recorded using the *OB-TraceVue* hospital signal using either ultrasound probes or scalp electrodes, and resampled into a beat-per-minute time series at 4 Hz. In this work we have decided to perform two basic transforms: FFT (Fast Fourier Transform) and CWT (Continuous Wavelet Transform) of both signals.

Note that for the visualization of FFT and CWT we have used the same (pathological) signal with pH $\leq$ 7.15 (no. 1001) from the DB. We did not perform any other filtering of the signal, nor performed any artifact removal (for example in the FHR signal, the drops to zero present an erroneous value and should be removed or replaced by some valid value).

Fast Fourier Transform. In Figs. 1 and 2 you can see the utilization of various window length of the FFT[3] for the FHR and UC signals respectively. As the sampling frequency remains the same, with changing the window size, we are not getting any new information. Therefore, in this work, wi did not follow this path and we performed the analysis using the CWT, providing us with different level of details.

In the referenced figures you can see that the accelerations and decelerations in FHR signal do not clearly correspond to any shapes in the spectrogram.

In the UC signal we can find patterns, related to increased uterine activity. Anyhow, the visual inspection of spectrograms does not bring much information (as opposed for example in the EEG signal).

Continuous Wavelet Transform. CWT (in our case Morlet wavelet [2] allows to specify one parameter $w0$ that changes the level of resolution in time vs. frequency space. In Figs. 3 and 4 you can see the varying results for FHR and UC signals respectively.

You can note that with changing the $w0$ parameter, the resolution (detail) in either frequency or time axis changes. Therefore we can obtain multiple details derived from one signal. Also, by using multiple details levels, we can increase the amount of training/testing samples, which is an interesting feature, as the most databases are low on the pathological instances.

---

[3] Short-Term FFT.

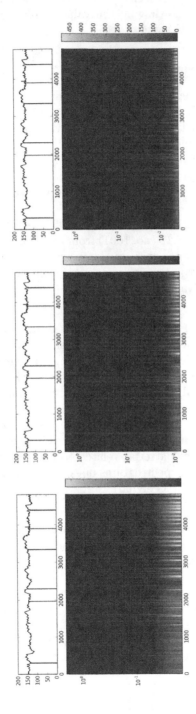

Fig. 1. Heart (FHR, pathological) signal after FFT for various windows length: (wlen param: 4, 8, 16). You can see that with varying window length the resolution is being lost and no new information is present. The vertical axis represents frequency (Hz), the horizontal represents time (s). The temperature values represent the energy in the respective frequency band.

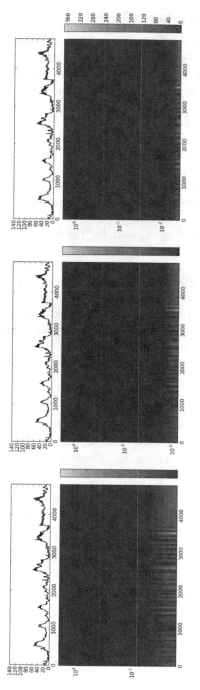

Fig. 2. Uterine (UC, pathological) signal after FFT for various windows length: (wlen param: 4, 8, 16). You can see that with varying window length the resolution is being lost and no new information is present. The vertical axis represents frequency (Hz), the horizontal represents time (s). The temperature values represent the energy in the respective frequency band.

3   Experimental Part

3.1   Methodology and Experiment Design

Data Preparation. In this work we have used the CNN networks for the task of classification of CTG recordings (with no clinical information added). As we intended to study the behavior of the networks for the task of signal classification only, we did not feed the network with any other information, although these are available in the CTG-UHB database.

In the data preprocessing stage, we have taken the last 14.400 samples[4] of the signals (both UC and FHR, sampled at 4 Hz) and performed a wavelet transform. We have used the Morlet wavelet (or Gabor wavelet) [2] which is composed of a complex exponential (carrier) multiplied by a Gaussian window (envelope). This wavelet is closely related to human perception, both hearing and vision. Using this approach we have obtained a 100×14400 gray-level pixel image that was fed to the convolutional CNN.

After the data have been preprocessed by the CWT with $w0$ params of $\{4, 8\}$, we have resampled the resulting 2D structure to small and large size with dimensions of 24×96 and 48×192 respectively. In the experiments we denote the two datasets with different resolution as *Small* and *Large*. From each recording, we have therefore obtained four 2D gray-level structures (two $w0$ details for both FHR, UC) in two different resolutions (small, large).

Each resolution (small or large) will be evaluated separately. This allows us to study the appropriate resolution that should be used in CTG signal pre-processing. The reason we have done this downsampling is to save the processor time.

The signals[5] (551 instances $\times 2$) were divided into two classes: normal ($pH > 7.15$, 439 instances) and pathological ($pH <= 7.15$, 112 instances). We have used the wavelet's $w0$ parameter values of 4 and 8 (this parameter controls the exchange of detail level in time or freq. domain). We have therefore obtained a dataset of 1756 and 448 class instances. These were randomly divided into training and testing set in a $2 : 1$ ratio.

Neural Network. The resulting 2D structures are fed to the convolutional network defined by the following parameters: K is the number of convolutional neurons in the first convolutional layer (with stride of 1). The next two convolutional layers contain $\frac{K}{2}$ and $\frac{K}{4}$ neurons (with stride of 2 for both). The number of neurons in the final (fully connected) layer is denoted N. In this layer, a dropout is performed with the parameter $pkeep$ which stands for $1 - dropout\_ratio$. The dimension of the convolutional filter is denoted $filter\_x$ and $filter\_y$.

---

[4] The common minimal lengtht of all recordings (lenght of the shortest recording). We are aware that this part of the signal might be affected by the decision of the physician and thus contain information about the outcome and subsequently bias this research.

[5] We have omitted the record no 4004 as it caused an unspecified error when processed by FFT.

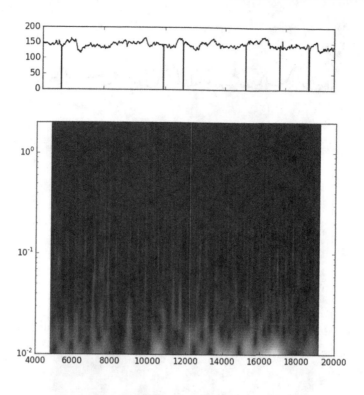

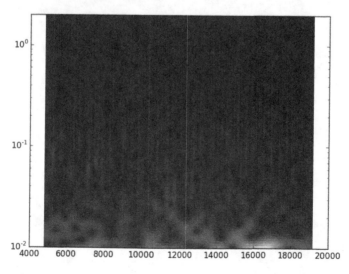

Fig. 3. Heart (FHR, pathological) signal (displayed at the top for reference) after CWT for various w0 params: (w0: 4, 8). The w0 parameter changes the detail ratio in time and frequency space. The vertical axis represents frequency (Hz), the horizontal represents time (samples). The temperature values represent the energy in the respective frequency band.

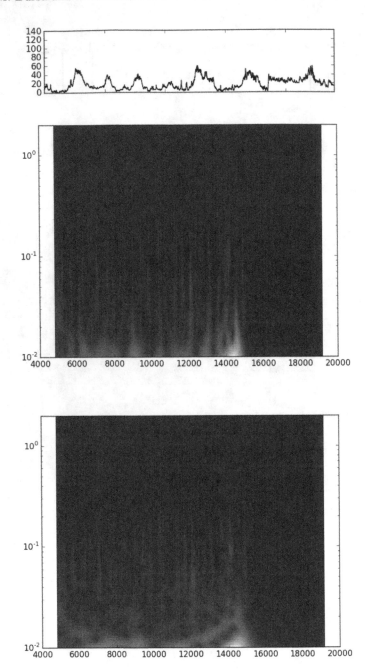

Fig. 4. Uterine (UC, pathological) signal (displayed at the top for reference) after CWT for various w0 params: (w0: 4, 8). The w0 parameter changes the detail ratio in time and frequency space. The vertical axis represents frequency (Hz), the horizontal represents time (samples). The temperature values represent the energy in the respective frequency band.

For the learning algorithm, a learning rate has to be defined. We have also used the (exponential) decay[6] to speed-up the learning at the beginning of the learning process. A number of iterations has also to be set.

As an optimization criterion (to be minimized), we have used Cross-entropy (binomial). It is defined as follows.

$$CE = -\frac{1}{n} \sum_{i=1}^{n} \sum_{j=1}^{m} y_{ij} \log(p_{ij}) \tag{1}$$

where i indexes samples/observations and j indexes classes; $y_{ij} \in \{0, 1\}$ is the sample (true) label, $p_{ij} \in (0, 1)$ is the prediction for the sample. The optimizer used was an Adam[7] first-order gradient-based optimizer [10].

Default Values. The default values of K were 5, for the N we have used the default value of 200. The default learning rate used was $lr = 0.003$ and default decay of $lrd = 0.95$. The dropout function used the parameter $pkeep = 0.75$ by default. The default convolutional filter dimensions were 8×16.

Some of the values have been modified during the experiments to study the behavior; if not, the above mentioned default values have been used.

3.2   Evaluation

We have performed multiple experiments (each batch has been run 10 times) to study the behavior of the network with various parameters. As a stopping criterion, the number of iterations (5000) has been used.

An example visualization of the learning process can be seen in the Fig. 5. The meaning of the legend is as follows: *tr.acc* and *te.acc* stand for training and testing accuracy (in %), *tr.ce* and *te.ce* stand for training and testing cross-entropy value. The *lrate* parameter is the learning rate (multiplied by 10000 for visualization purposes). As a stopping parameter we have used the number of iterations (5000), the stopping criterion used might also be the drop in cross-entropy.

Neural Network Layer. The main parameters of the network are the number of neurons in the first convolutional layer K and number for neurons in the fully connecterd network N. We have used varying values of the parameters.

First, we have studied the behavior of the K parameter that did not show any distinct trend by itself (see Fig. 6). On the other hand, for the N parameter we can clearly see a trend that peaks at about 200 neurons for both large and small dataset (see Fig. 7). Note that in this case it is a summary of the larger experiment, therefore the other value (i.e. K when varying N) was not fixed. This might affect the plots. However, the behavior is detailed in the following paragraph.

---

[6] `tf.train.exponential_decay()`.
[7] `tf.train.AdamOptimizer()`.

Visualization of the learning process

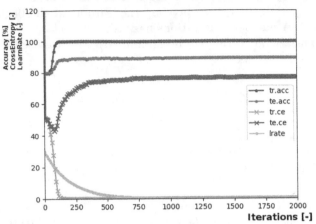

Fig. 5. Accuracy [%] (`tr.acc`, `te.acc`), cross-entropy [–] (`tr.ce`, `te.ce`) and learning rate [–] (`lrate`) for both testing (`te`) and training (`tr`) sets. Note that for visualization purposes, the learning rate has been multiplied by 10.000 and number of iterations has been se to 2000 only (in real experiments, the limit was 5000). You can notice a drop (lower extrema) in the testing cross-entropy. This can also be used as stopping criterion.

In Figs. 8 and 9 we detailed the behavior of various (K, N) values. We can observe similar findings as mentioned above: the extreme for N at about the value of 200, no visible pattern for the K parameter.

We have also studied various convolutional filter dimensions. Again, no clear trends were observed in the results. However, if we have evaluated (grouped) the results by the filter size $(fx * fy)$, the best values were 16×16 for small and 16×12 for large dataset.

Learning Process. In Figs. 10 and 11 you can see the results for various values of learning rate and (exponential) learning rate decay. In case of the learning rate we can estimate that the optimal values for the small (small resolution) dataset are located around 0.0005 and 0.003, for the large (large resolution) the values are most likely to be slightly shifted to values between 0.0002 and 0.003. In this range we can notify that the variability of the results lower and is located in higher mean values of the accuracy parameter.

For the learning rate decay, we did not find any obvious pattern in the results. This might be caused by the stochastic nature of the experiments. We therefore need to run more experiments to evaluate this parameter of the learning process. Also, an visual inspection can be used to fine-tune the parameter.

The same holds also for the dropout ratio (the *pkeep* parameter).

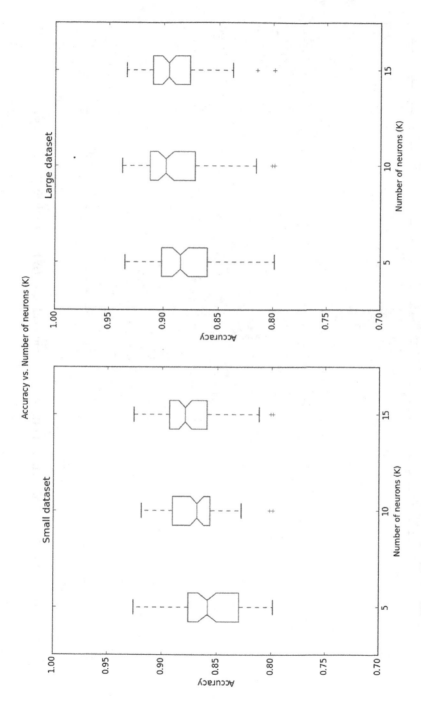

Fig. 6. Number of the neurons in the first convolutional layer (K) for small and large datasets.

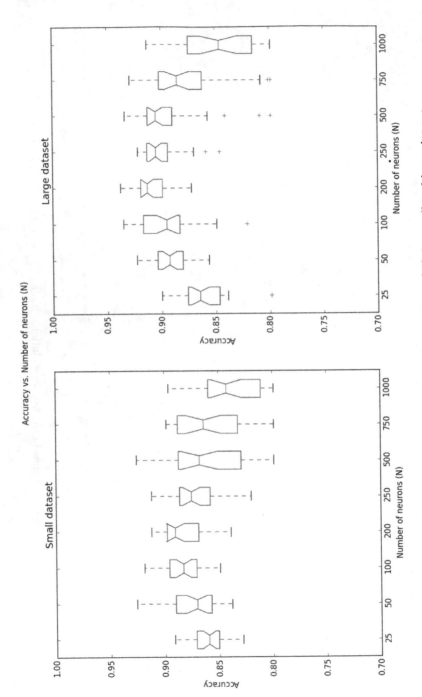

Fig. 7. Number of the neurons in the non-convolutional layer (*N*) for small and large datasets.

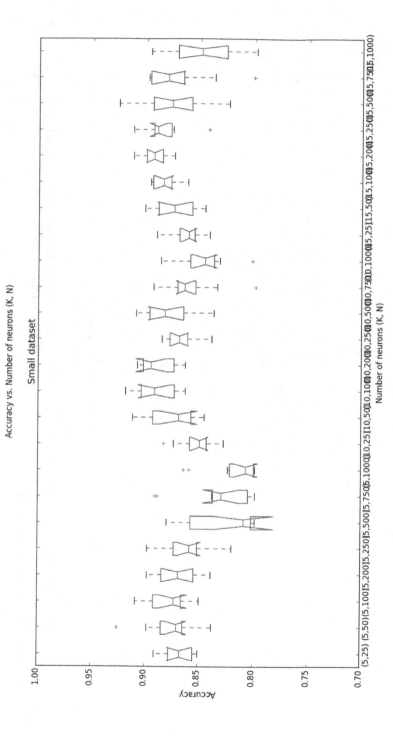

Fig. 8. Various combinations of neuron count in the convolutional (K) and non-convolutional layers (N) for small dataset. The values for K are: {5, 10, 15}, for N: {25, 50, 100, 200, 250, 500, 750, 1000}. The horizontal axis displays the combinations. You can notice that the results are mainly affected by the N value.

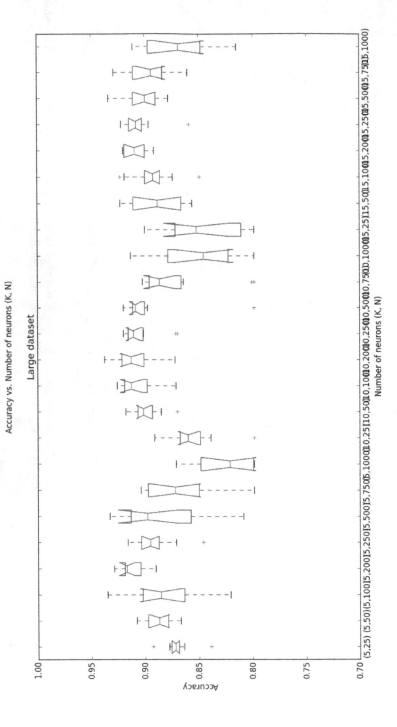

Fig. 9. Various combinations of neuron count in the convolutional (K) and non-convolutional layers (N) for large dataset. The values for K are: {5, 10, 15}, for N: {25, 50, 100, 200, 250, 500, 750, 1000}. The horizontal axis displays the combinations. You can notice that the results are mainly affected by the N value.

4   Conclusion and Final Results

The best accuracy obtained so far was 0.941, reached after only 2000 iterations for the network with the following setup: 3 convolutional layers with (5, 10, 20) output neurons ($K = 5$) and respective filter size (16×8, 8×4, 8×4), filter strides (1, 2, 2). The number of neurons in fully connected layer was $N = 200$. We have used the *relu* activation function, and dropout ratio of 0.25. The starting learning rate was 0.003 with learning rate deacy of 0.95.

The optimizer used was an Adam first-order gradient-based optimizer [10] with exponential learning rate decay. We have optimized the minimal cross-entropy value.

We have also studied various organization structures of the network, different subsampling parameters of the processed images and various parameters. The results are sensitive to the value of N and to some extent to learning rate and filter sizes, although for the latter we did not find any clear values.

As we have obtained promising results, in the future work we are going to improve the experimental and evaluation process (crossvalidation, sensitivity/specificity) and combine the various 2D structures into one (as in RGBA) image. We also need to find out which signal the network prefers for the final decision, or, how it combines the signals to achieve the classification decision (cf. Figs. 8 and 9).

Acknowledgment. The research is supported by the project No. 15-31398A Features of Electromechanical Dyssynchrony that Predict Effect of Cardiac Resynchronization Therapy of the Agency for Health Care Research of the Czech Republic. This work has been developed in the BEAT research group https://www.ciirc.cvut.cz/research/beat with the support of University Hospital in Brno http://www.fnbrno.cz/en/. Access to computing and storage facilities owned by parties and projects contributing to the National Grid Infrastructure MetaCentrum provided under the programme *Projects of Large Research, Development, and Innovations Infrastructures* (CESNET LM2015042), is greatly appreciated.

Appendix: Results

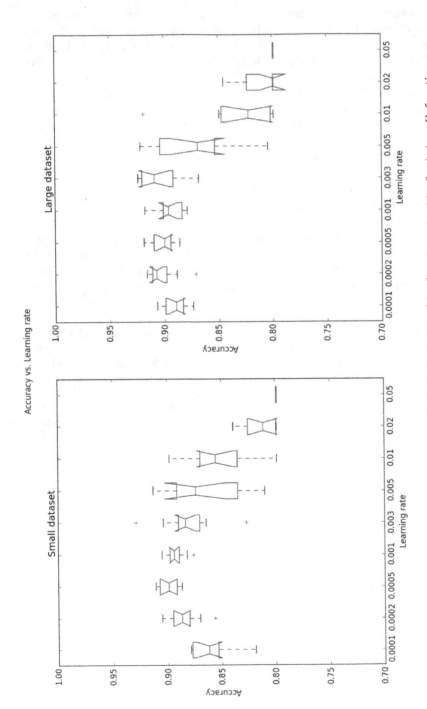

Fig. 10. Learning rate for small and large datasets. Used in the `tf.train.AdamOptimizer()` function.

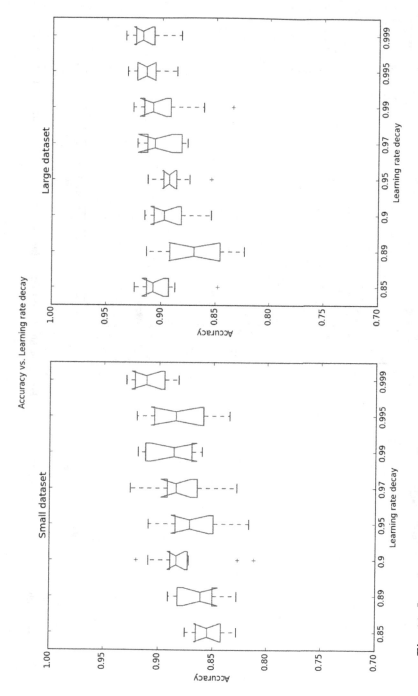

Fig. 11. Learning rate decay for small and large datasets. Used in the `tf.train.exponential_decay()` function.

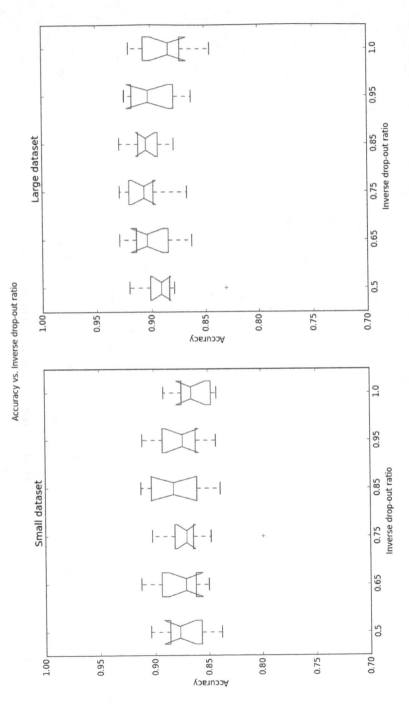

Fig. 12. Dropout parameter (pkeep, number of neurons kept) for small and large datasets. Used in the tf.nn.dropout() function.

References

1. Abadi, M., Agarwal, A., Barham, P., Brevdo, E., Chen, Z., Citro, C., Corrado, G.S., Davis, A., Dean, J., Devin, M., Ghemawat, S., Goodfellow, I., Harp, A., Irving, G., Isard, M., Jia, Y., Jozefowicz, R., Kaiser, L., Kudlur, M., Levenberg, J., Mané, D., Monga, R., Moore, S., Murray, D., Olah, C., Schuster, M., Shlens, J., Steiner, B., Sutskever, I., Talwar, K., Tucker, P., Vanhoucke, V., Vasudevan, V., Viégas, F., Vinyals, O., Warden, P., Wattenberg, M., Wicke, M., Yu, Y., Zheng, X.: TensorFlow: Large-scale machine learning on heterogeneous systems (2015). http://tensorflow.org/, softwareavailablefromtensorflow.org
2. Bernardino, A., Santos-Victor, J.: A real-time gabor primal sketch for visual attention. In: Marques, J.S., Pérez de la Blanca, N., Pina, P. (eds.) IbPRIA 2005. LNCS, vol. 3522, pp. 335–342. Springer, Heidelberg (2005). doi:10.1007/11492429_41
3. Bursa, M., Lhotska, L., Chudacek, V., Spilka, J., Janku, P., Hruban, L.: Information retrieval from hospital information system: Increasing effectivity using swarm intelligence. J. Appl. Logic 13(2, Pt. A), 126–137 (2015). http://www.sciencedirect.com/science/article/pii/S1570868314000809, sI: SOCO12
4. Chudacek, V., Spilka, J., Bursa, M., Janku, P., Hruban, L., Huptych, M., Lhotska, L.: Open access intrapartum ctg database. BMC Pregnancy Childbirth 14, 16 (2014)
5. Chudáček, V., Spilka, J., Huptych, M., Lhotská, L.: Linear and non-linear features for intrapartum cardiotocography evaluation. Computing in Cardiology 2010 Preprints. IEEE, New Jersey (2015)
6. Doret, M., Spilka, J., Chudáček, V., Gonçalves, P., Abry, P.: Fractal Analysis and Hurst Parameter for intrapartum fetal heart rate variability analysis: A versatile alternative to Frequency bands and LF/HF ratio. PLoS ONE 10(8), e0136661 (2015). http://dx.doi.org/10.1371%2Fjournal.pone.0136661
7. Goldberger, A.L., Amaral, L.A.N., Glass, L., Hausdorff, J.M., Ivanov, P.C., Mark, R.G., Mietus, J.E., Moody, G.B., Peng, C.K., Stanley, H.E.: PhysioBank, PhysioToolkit, and PhysioNet: Components of a new research resource for complex physiologic signals. Circulation 101(23), e215–e220 (2000). Circulation Electronic Pages: http://circ.ahajournals.org/cgi/content/full/101/23/e215
8. Hruban, L., Spilka, J., Chudáček, V., Janků, P., Huptych, M., Burša, M., Hudec, A., Kacerovský, M., Koucký, M., Procházka, M., Korečko, V., Seget'a, J., Šimetka, O., Mchurová, A., Lhotská, L.: Agreement on intrapartum cardiotocogram recordings between expert obstetricians. J. Eval. Clin. Pract., May 2015. http://dx.doi.org/10.1111/jep.12368
9. Huser, M., Janku, P., Hudecek, R., Zbozinkova, Z., Bursa, M., Unzeitig, V., Ventruba, P.: Pelvic floor dysfunction after vaginal and cesarean delivery among singleton primiparas. Int. J. Gynecol. Obstet. 137(2), 170–173 (2017). http://dx.doi.org/10.1002/ijgo.12116
10. Kingma, D.P., Ba, J.: Adam: A method for stochastic optimization. CoRR abs/1412.6980 (2014). http://arxiv.org/abs/1412.6980
11. Spilka, J., Frecon, J., Leonarduzzi, R., Pustelnik, N., Abry, P., Doret, M.: Sparse support vector machine for intrapartum fetal heart rate classification. IEEE J. Biomed. Health Inform. 21(3), 664–671 (2017)
12. Spilka, J., Chudáček, V., Janků, P., Hruban, L., Burša, M., Huptych, M., Zach, L., Lhotská, L.: Analysis of obstetricians decision making on CTG recordings. J. Biomed. Inform. 51(0), 72–79 (2014). http://www.sciencedirect.com/science/article/pii/S1532046414000951

Reducing Red Blood Cell Transfusions

James P. McGlothlin[1]([✉]), Evan Crawford[1], Hari Srinivasan[1],
Carissa Cianci[2], Brenda Bruneau[2], and Ihab Dorotta[2]

[1] Fusion Consulting Inc., Irving, TX, USA
jmcglothlin@fusionconsultinginc.com
[2] Loma Linda University Health System, Loma Linda, CA, USA

Abstract. The health care industry in the United States is undergoing a paradigm shift from the traditional fee-for-service model to various payment and incentive models based on quality of care rather than quantity of services. One specific scenario where more treatment does not equate to better care is red blood cell transfusions. While blood transfusions often save lives, there are numerous complications which can result and blood should be transfused only if medically necessary. Several studies have indicated that a very high percentage of units transfused are not clinically appropriate. These transfusions increase cost and negatively impact patient outcomes. In this paper, we will present an analytics project to identify and track the transfusions which are performed without clear necessity. Furthermore, we will describe how we utilized data discovery and supervised learning to improve our classification algorithm and the accuracy of our results. We will demonstrate that our project is effectively reducing red blood cell transfusions.

Keywords: Healthcare analytics · Data warehouses · Blood utilization · Quality

1 Introduction

Traditionally, providers in the United States, including physicians and hospitals, have been reimbursed for the procedures they perform and the services they provide. This model is known as fee-for-service. Many healthcare experts and economists have concluded that this payment system is inefficient and unsustainable because it rewards excessive, expensive and sometimes unnecessary medical treatment. For example, under a fee-for-service model, a hospital is actually paid more money if the patient stays in the hospital longer. However, a lower length of stay where the patient is able to return to their home is considered a better outcome. Numerous new models have been introduced over the last 10 years in an attempt to encourage efficient care and reward quality outcomes. These include Pay for Performance, Value-based Purchasing, Merit-based Incentive Payment System (MIPS), Medicare Stars, Advanced Alternative Payment Models (APMs), Medicare Shared Saving Program (MSSP) and more. The idea behind these programs is to reimburse providers based on diseases, diagnosis-related groups (DRGs), quality metrics and case mix index rather than on actual length of stay or procedures performed. Then, provide incentives and rewards to the providers for good outcomes, and sometimes penalties for poor outcomes.

© Springer International Publishing AG 2017
M. Bursa et al. (Eds.): ITBAM 2017, LNCS 10443, pp. 120–133, 2017.
DOI: 10.1007/978-3-319-64265-9_10

Red blood cell transfusions are one of the most common clinical procedures in an acute hospital. One in seven hospitalized patients will receive at least one unit of blood [4]. Historically, 50% of all patient admitted to the ICU receive a blood transfusion [1] and the average ICU patient receives 5 units [2]. 36,000 units of blood are transferred every single day in the United States alone [3]. Additionally, many other cohorts have high transfusion rates including over 20% of hip replacement patients [5] and almost 50% of coronary artery bypass graft (CABG) patients [47]. Most studies show that blood transfusions are ordered inappropriately more than 20% of the time with several studies estimating the rate at over 50% [8, 9, 39, 42]. Thus, there is opportunity to significantly reduce blood transfusions.

As the next section clearly documents, there are many adverse complications directly indicated by red blood cell transfusions. As a result, patient outcomes, including hospital-acquired infections, length of stay, readmissions and mortality, are all negatively impacted by red blood cell transfusions. Furthermore, blood transfusions have a huge cost factor and are almost always performed at a loss. The cost to acquire a unit of blood is $278 but the estimated cost to test, store, transport and administer the blood brings the total cost to approximately $1,100 [10]. This does not even taken into account the cost of negative outcomes and increased length of stay. Average reimbursement for a unit of blood is only $180 creating a significant financial loss to the hospital [10].

There are prior studies demonstrating the potential opportunities associated with changing physician behavior related to ordering of blood transfusions. Stanford University was able to reduce blood transfusions 24% and generate savings of $1.6 million a year simply by requiring physicians to review guidelines prior to signing transfusion orders [6]. For transfusion patients, the mortality rate decreased from 5.5% to 3.3%, and the length of stay decreased from 10.1 days to 6.2 days [5, 6]. In a 2012 study analyzing clinical data from 464 hospitals, it was asserted that 802,716 units of blood could be saved annually which would equate to a savings of $165 million [7]. In another study at an Australian acute care facility, transfusions were reduced 28.8% by standardizing physician policies [43].

In this paper, we will outline our analytics solution for identifying and reducing inappropriate red blood cell transfusions. First, we will survey the clinical complications and risks associated with blood transfusions. Then, we will review the literature for appropriate transfusion criteria and define our parameters. Next, we will define our technical implementation and the challenges we faced. Next, we will describe our analytics solutions and how it is used by the clinicians. Then, we will describe our approach for supervised learning and data discovery and how we used it to alter our classification algorithm. We will go over results from our project so far. Finally, we will propose future work and make some conclusions.

2 Problem Domain

To understand some of the ramifications of unwarranted red blood cell transfusions we need to highlight some of the adverse complications that can occur. In a survey of surgeons who administered blood despite high hemoglobin, the number one answer was "to be safe" [11]. However, blood transfusions are never safe. Red blood cell transfusions weaken the immune system and thus have been shown to increase infection rates, organ failure progression and even septic shock. Common adverse reactions can occur with each administration of blood product include anaphylactic (allergic) reactions (3 in 100 patients) [18], transfusion induced fever [21], and iron overloads resulting in hemochromatosis [23].

Transfusion-related acute lung injury (TRALI) is a known risk which can cause acute respiratory distress syndrome and even death [19, 24]. "In recent years, transfusion-related acute lung injury (TRALI) has developed from an almost unknown transfusion reaction to the most common cause of transfusion-related major morbidities and fatalities" [25].

An acute hemolytic transfusion reaction (ABO incompatibility) occurs when the red blood cells that were given during the transfusion are destroyed by the person's immune system [26, 27]. This usually happens rapidly after the transfusion but the reaction can be delayed for patients with specific co-morbidities including sickle cell [28, 29]. Acute hemolytic transfusion reaction is a life-threatening event.

Febrile non-hemolytic transfusion reactions (FNHTR) is a reaction to transfusion generally marked by fever. FNHTR is usually caused by antibodies directed against donor leukocytes and HLA antigens [31, 33]. While FNHTR is often non-severe, it can have severe impact especially on the elderly [35].

Additional risks associated with transfusions include anaphylaxis, contraction of blood borne illnesses [30], post-transfusion purpura (PTP) (severe immune thrombocytopenia) [36, 37] and graft vs host disease (GvHD) [34].

Numerous studies have demonstrated the impact of blood transfusions on patient outcomes. In a 12 year study that ended in 2000, it was determined that patient who receive blood transfusions postoperatively are at a significant risk of developing a postoperative bacterial infection due to immunosuppression [38]. For hip replacement patients, "blood transfusion was the most important predictor of discharge around the third day of admission, as patients receiving blood had a three-fold increased risk of staying more than 3 days" [5, 40]. One study of 9,245 total knee and hip replacement patients showed that "the quantity of allogeneic blood units that was transfused was an independent predictor of PJI (peri-prosthetic joint infection) after primary joint arthroplasty" [5, 41]. In a study of 285 intensive care units (ICUs) at 213 hospitals, a 2001 study showed that the number of red blood cell transfusions a patient received is independently associated with longer ICU length of stay, longer total length of stay and increased mortality rate [44]. Blood transfusions during trauma and surgery is "independent predictor of multiple organ failure, systemic inflammatory response syndrome, increased infection, and increased mortality" [46]. In another study specific to coronary artery bypass graft (CABG) patients, a study of 11,963 such patients, "perioperative red blood cell transfusion is the single factor most reliably associated with increased risk of postoperative morbid events. Each unit of red cells transfused is associated with incrementally increased risk for adverse outcome" [47]. In another study of 2,085 critically ill patients, the number of transfusions was independently associated with nosocomial infection [50]. "When corrected for survival probability, the risk of nosocomial infection associated with red blood cell transfusions remained statistically significant ($p < .0001$). Leukoreduction tended to reduce the nosocomial infection rate but not significantly. Mortality and length of stay (intensive care unit and hospital) were significantly higher in transfused patients, even when corrected for illness severity" [50].

There are significant complications and safety risks associated with red blood cell transfusions across multiple patient cohorts, as demonstrated by numerous studies.

3   Classification

We have demonstrated the need to limit blood transfusions when not clinically required. Next, our challenge is to identify when blood transfusions are clinically appropriate. There is no doubt that red blood cell transfusions are appropriate and save lives. It is estimated that 4.5 million Americans would die each year if not for blood transfusions [48]. Red blood cell transfusions are indicated when necessary to facilitate the rapid increase of oxygen supply delivered to tissues. The challenge is the only effective measure to evaluate this oxygen supply is intracellular pO2 however this parameter is not accessible and usable in a clinical setting, so we are left to rely on surrogate indicators, primarily hemoglobin (Hb) Hb and the hematocrit (Htc) [49].

For many years, the standard criteria for blood transfusion was Hg/dL <10 [51]. Now, it is generally agreed that blood transfusions are required when Hg/dL <7 or there is a estimated blood loss of 40% of the patient's volume [49]. It is still agreed that Hg/gL >10 is contraindicative of the need for a transfusion. The challenge is patients with Hg between 7 and 10 and/or estimated blood loss between 20 and 40%. In [49], a variety of factors are proposed for the physician's consideration, but our goal is to try to classify the appropriateness based entirely on discrete clinical indicators. We understand that clinicians will still have to make the decision, but if we can classify transfusions we can suggest guidelines and monitor progress.

We studied many proposed guidelines including [1, 6, 8, 9, 49, 13] and we worked with clinical leadership and decided the following criteria to clinically indicate blood transfusion for a patient with Hg >7 g/dL:

1. Hemodynamic instability
2. Evidence of severe bleeding
3. Evidence of acute myocardial infarction (AMI).

For our first pass, we used the following criteria to classify blood transfusion orders:

- Category 1: Hemoglobin <7.0 g/dL in a hemodynamically stable patient (systolic blood pressure >90 mmhg
- Category 2: Hemoglobin <8.5 g/dL and one of the following (as indicators of severe bleeding)
  - Drop of hemoglobin by at least 1.5 g/dL in previous 24 h
  - Documented blood loss
  - Systolic blood pressure <90

- Category 3: Hemoglobin <8.5 g/dL with hemodynamic instability, indicated by one of the following:
  - Lactate >3
  - Base deficit >5
  - ScvO2 <70%
- Category 4: Hemoglobin <10 g/dL and evidence of AMI (acute myocardial ischemia or infarction)
- Category 5: All other clinical reasons (Did not meet category 1–4 guidelines)

This classification is done per transfusion order, and thus based on the order time. So a patient may qualify at other points during his or her encounter but still not qualify at the time of the order.

We state that this is our phase 1 criteria because this is not the classification algorithm we are currently using. Later in this paper, we will describe how this criteria has been updated.

4 Implementation Details

We have designed this as an analytics and business intelligence project which will leverage our enterprise data warehouse. Our goal is to identify the follow data points:

1. Red blood cell transfusions ordered (including time of the order, who made the order, where the order was made, and how many units were ordered)
2. The patient's appropriateness classification at the time of the order as defined by the indicators above
3. For each unit, if the unit was administered and the time of the administration.

This will provide us two separate but related measures, the number of units ordered, by appropriateness classification, and the number of units administered by appropriateness classification. Our goal is to reduce four related but distinct data points:

1. the overall number of units ordered (both appropriate and inappropriate)
2. the number of units ordered which are classified as inappropriate
3. the overall number of units administered
4. the number of units administered which are classified as inappropriate.

Obviously, #4 is the most important key performance indicator (KPI) as it will tell us whether the patient received inappropriate blood product and thus was potentially subject to unnecessary risk and complication. However, we are hoping that our clinicians will even be able to reduce the number of appropriate orders by attempting other less invasive clinical interventions. These can include [52]:

- medications to treat anemia
  - ferrous sulphate
  - folic acid
  - B12 vitamin
  - erythropoietin

- darbepoetin
- continuous erythropoietin activator (CEA)
• Medications to prevent blood loss
  - Tranexamic acid
  - aminocaproic epsilon acid
  - vasopressin
  - combined estrogens
  - octreotide
  - somatostatin
  - desmopressin acetate (DDAVP)
  - K vitamin (phytomenadione)
  - activated recombined factor VII
  - factor VIII coagulation concentrate
  - prothrombin complex concentrate, human
  - fibrinogen concentrate
  - human recombined factor XIII
• Topical dressings, medications and ointments to reduce bleeding
• Equipment which reduces blood loss and saves patient's own blood

Additionally, we want to track ordered blood, even if the blood was not ultimately given. When blood is ordered, it must be appropriately typed, transported and made available for administration. Thus, costs and inventory challenges are created when blood is ordered even if the unit is never administered.

Nonetheless, our most important metric is the number of units administered, especially the number of units which were not clinically required.

4.1   Data Implementation

Loma Linda University Health System (LLUHS) has an enterprise data warehouse (EDW) that leverages the data warehouse provided by its EMR (electronic medical record system) but which has been extended and customized to address the analytic needs and opportunities of Loma Linda. In this data warehouse, we have a fact table which includes every procedure order entered into the system. This allows us to easily identify each time a provider orders a blood transfusion.

We then look back from that time at two additional fact tables to analyze the clinical appropriateness of the order. These fact tables are diagnostic lab results and "flowsheets", a generic term for clinical documentation including vital signs. This was mostly straight forward with 2 exceptions. We want to know if the hemoglobin dropped 1.5 in a 24 h period. Since an arbitrary number of Hg results could be taken in any given 24 h, since this is a moving 24 h period not a calendar day, and since we want the delta between the max and the minimum rather than adjacent reductions, this is too complicated a query for us to perform at report time. Therefore, we created a special fact table, at the grain of lab result, which shows for that result the delta between that result and the maximum value for that diagnostic indicator in the previous 24 h. This provides a much simpler table to query during our classification algorithm.

The other clinical indicator that was challenging was documented blood loss. For our first phase, we made the assumption that any surgical procedure incurred blood loss, and that furthermore any documentation indicating an estimated blood loss should be included regardless of volume. As we stated before, this algorithm is improved in phase 2 documented later in this paper.

Finally, the last remaining data point we need is the number of units ordered, whether they were given and when they were given. We are classifying the appropriateness of each unit based on the time it was ordered not the time it was administered. However, we want the time it was administered because we are only looking for our clinical indicators between the time of an order and the time of the previous administration.

This data point proved extremely challenging. We tried to use the "input/output" documentation that indicates what products are put into a patient. However, this workflow did not distinguish red blood cells from other products such as frozen plasma. Our operational data store has a table specifically dedicated to blood administration. However, this table was only populated when the provider scanned the unit prior to administration, which we found was not done consistently. We tried utilizing the blood transfusions coded on the hospital bill but we were unable to tie them back to specific orders. We tried using the medication administration record (MAR) but found this was also inconsistent.

Our EMR has a concept of parent and child orders. A parent order for 4 units of red blood cells will create 4 child orders. This information was not in our data warehouse so we created an extension to make it available. We used the child order release time as the blood transfusion time. We used the child order status to indicate whether the blood unit was administered or not. Once again, this logic is updated in our second phase, detailed later in this paper.

4.2 User Implementation

We used our data to create 2 dashboards. The primary dashboard shows volume of units ordered and administered by appropriateness category, and the trend of what percentage of the orders were appropriate. It can be filtered to specific departments, services, providers, diagnoses, DRGs, locations and dates. However, the rules for how the units are counted and classified cannot be changed. An example of this dashboard is shown as Fig. 1.

One of the most important features of our dashboards is it allows the user to drill down to the low level details. So for example, for every inappropriate blood transfusions, the user can drill down and get the order time, the provider, the patient information, the order identifier, etc. Our experience is providing this level of details increases the trust clinicians have in our analytics dashboards.

Additionally, we created another dashboard that is a data discovery version of our primary dashboard. This dashboard allows the user to see details about cancelled units and optionally even count these units as completed. This dashboard also lets the user utilize rules to switch options such as the Hg, lactate or blood pressure thresholds used in classification. This allows the user to perform "what-if" scenario evaluations. Our

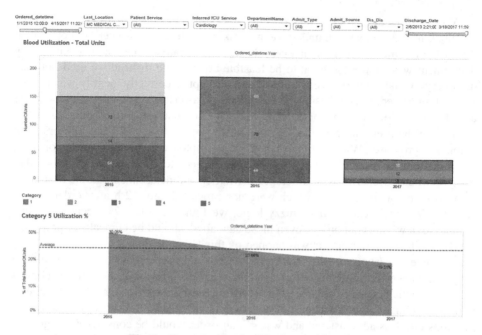

Fig. 1. Example blood utilization dashboard

intention is not that the end user utilize this dashboard but rather that the performance improvement experts in Loma Linda's Patient Safety and Reliability department utilize these options to learn more about "why" results occurred and if appropriate to adjust the logic. We will go over how we utilized data discovery in the next section.

5 Analyzing and Improving the Algorithm

5.1 Changes from User Acceptance Testing

Once our first phase was done, we performed a rigorous user acceptance testing (UAT). Not only did we have clinicians compare our results with the patient charts, but we even completed a one month exercise where we had a nurse practitioner in the medical intensive care unit (MICU) keep manual records of all transfusions. While our data was extracted correctly according to the guidelines chosen in Sects. 3 and 4, we found there were significant issues with the quality of the source data.

One of the biggest issues we discovered is that the status on the child order was not trustworthy. Over 30% of child orders for red blood cell units had a status of cancelled even though our manual review indicated the units were actually administered. This is a process problem, not all clinicians were resulting the order prior to administration. Our EMR automatically changes the status of any order which is not resulted to cancelled upon transfer or discharge. Thus we have orders which were performed by the system has recorded the status as cancelled.

We considered improving the workflow and clinician education to correct this documentation, and we may still undertake this task in the future. However, we determined the task of retraining nurses in every ICU, perioperative location and emergency department was simply too great to be feasible at this time. Instead we used our data discovery to analyze the cancelled orders and attempt to correctly identify which units were administered. We tried using the cancellation reason hoping that this could separate actual cancellation from automated ones, but this did not achieve improved results. We tried using the user who cancelled it, once again hoping to separate actual users from automated processes. When all of this failed, we attempted to utilize the cancellation time, since automated cancellations would not happen immediately. We used our dashboard to analyze what-if scenarios with different timeframes and determined that the best algorithm was to classify all units which were not cancelled until 12 h after order as performed. While we realize this is fuzzy logic, we found it to be 97% accurate against manual review. Since all orders, even those cancelled, should be clinically appropriate, we did not feel that our project was harmed by this slight deviation.

However, because so many child orders were not properly resulted, we could no longer trust the unit administration time. In our first phase, we were using clinical indicators which occurred between the order time and the previous transfusion time to determine clinical appropriateness. For example, if the Hg was measured after the previous unit was administered and was <7, an order would be considered category 1. If we do not accurately know when the previous unit was administered, this algorithm does not work. Therefore, we converted the algorithm to look back in the 24 h prior the order, and not be dependent on the previous transfusion time.

This concludes the updates we had to make as a result of validation, but we still want to determine if our classification can be improved.

5.2   Supervised Learning

Our projection is a classic classification problem, classifying each order into a bucket, or appropriateness category. Our preset phase 1 algorithm is a starting place in that it classified each order. The question is, were those classifications correct? If we can determine, manually or automatically, the true final classification of each order we can adjust the algorithm using well-known machine learning techniques for classifications.

First, we wanted to determine for each order that was classified as inappropriate whether it was truly not clinically justified. We did not want the ordering clinician to make this classification decision since they have natural bias (they already ordered the unit so they will want to justify it). Therefore, we utilized the MICU NP from our user acceptance testing to also indicate appropriateness for each category 5 order.

This gives us a partial set of classified outcomes to retrain our algorithm on. However, there is a much larger set of data available for reclassification in the reverse. Each time a patient qualified for one of the four categories but had no blood ordered, this is a negative reclassification indicating the clinician decided that blood was not clinically necessary. The advantage of this approach is this data does not require manual review and input.

We collected our reclassification data and then had our statistics department analyze it using two well-known machine learning methodologies for supervised learning: support vector machines [53, 54] and random forests [55]. We limited to the analysis to the data points we were already using: Hg/dL, Hg delta over 24 h, lactate, estimated blood loss, whether a surgery was performed, systolic blood pressure, base deficit, ScvO2 and presence of AMI or other diagnosis. We then took the results from the machine learning and rounded them because we felt there was no advantage to asking a provider to follow a procedure with a confused exact number such as Hg <6.91. We then reviewed the results with a clinician and came up with the following changes:

- Hemoglobin drop within 24 h threshold increased from 1.5 to 2
- New threshold for blood loss of 300 ml
- Hemoglobin <8.5 and presence of AMI, rather than <10
- Include diagnosis of chest pain or acute coronary syndrome in addition to AMI

We were pleased to find all of our changes were more restrictive rather than less, supporting our overall goal of having a restrictive but safe blood transfusion policy.

5.3   Phase Two

We reimplemented our data warehouse and dashboard as follows:

- Count all child orders which are completed or are not cancelled until >12 h after order time as a unit of blood administered
- Switch all look back times for clinical indicators to 24 h
- Change the classification algorithm as follows:
  - Category 1: Hemoglobin <7.0 g/dL in a hemodynamically stable patient (systolic blood pressure >90 mmhg)
  - Category 2: Hemoglobin <8.5 g/dL and one of the following (as indicators of severe bleeding)
      Drop of hemoglobin by at least 2 g/dL in previous 24 h
      Documented blood loss greater than 300 ml
      Systolic blood pressure <90
  - Category 3: Hemoglobin <8.5 g/dL with hemodynamic instability, indicated by one of the following:
    - Lactate >3
    - Base deficit >5
    - ScvO2 <70%
  - Hemoglobin <8.5 g/dL and evidence of AMI, acute coronary syndrome or diagnosis of chest pain

This is the algorithm we are currently using for our project and to generate the results indicated in this paper.

6 Results

Across the hospital organization, we have seen a reduction of red blood cell units given by 5.6% since the start of our analytics program. Specifically, inappropriate transfusions have dropped by 10.2%. However, the rate of inappropriate transfusions remains high (35.8%).

However, our initial rollout was concentrated on the adult ICU. Only recently have we begun to look at perioperative services, the emergency department and pediatrics. In the adult ICU, our inappropriate transfusion rate has been reduced to 26.0% and our number of units administered has been reduced by 49.8% over the last 15 months.

have seen an overall reduction of units administered over the last 6 months of 1626 units. Using the cost numbers from [10] this has the potential to generate annual acquisition cost savings of $900,000 and complete savings of $3.57 million.

7 Future Work

For future work, we plan to adjust our estimated blood loss algorithm to take into account patient blood volume. This is necessary for that category to be effective for pediatric patients. We also would like to separate clinically necessary vs "borderline" patients so we could also target reductions in patients for whom alternative therapies may be possible. Furthermore, we want to analyze overuse of transfusions even when a transfusion is appropriate (in other words, whether more units were given than necessary). We plan to analyze our results for statistical significance and compare them to length of stay, readmissions and mortality to monitor our program's effective on our patient's outcomes. We also plan to work with our finance and decision support teams to acquire a more precise savings estimate rather than relying on data from literature. Finally, we may attempt to retrain our classifier algorithm again to further improve results. We are also going to utilize a similar approach to look at other utilization outside of blood, such as MRI imaging.

8 Conclusions

The preponderance of literature and research studies demonstrates that red blood cell transfusions are often over prescribing, and that transfusions negatively impact patient health. We have created an analytics solution which includes a dashboard to provide insight into clinically unwarranted blood transfusions. We have described how we built this tool and how we have adjusted our algorithm from optimal and efficient classification of transfusion orders. Finally, we have demonstrated our early results, which show that this project has effectively reduced unnecessary blood transfusions and reduced cost at Loma Linda University Health System.

References

1. Littenberg, B., Corwin, H., Leichter, J., AuBuchon, J.: A practice guideline and decision aid for blood transfusion. Immunohematology **11**, 88–94 (1995)
2. Corwin, H.: Anemia and blood transfusion in the critically ill patient: role of erythropoietin. Crit. Care **8** (2004)
3. American Association of Blood Banks (2013)
4. Hemez, C.: Blood Transfusion Costs. Yale Global Health Review (2016)
5. Lee, J., Han, S.: Patient blood management in hip replacement arthroplasty. Hip Pelvis. **27**(4), 201–208 (2015)
6. Anthes, E.: Evidence-based medicine: save blood, save lives. Nature **520**(7545), 24–26 (2015)
7. Brimmer, K.: Premier: tech could help save millions on blood transfusions. Healthcare IT News (2012)
8. Mozes, B., Epstein, M., Ben-Bassat, I., Modan, B., Halkin, H.: Evaluation of the appropriateness of blood and blood product transfusion using preset criteria. Transfusion **29**, 473–476 (1989)
9. Ghali, W., Palepu, A., Paterson, W.: Evaluation of red blood cell transfusion practices with the use of preset criteria. Can. Med. Assoc. J. **150**, 1449–1454 (1994)
10. Khan, A.: Blood Transfusions Overused, Study. International Business Times (2012)
11. Hébert, P., Schweitzer, I., Calder, L., Blajchman, M., Giulivi, A.: Review of the clinical practice literature on allogeneic red blood cell transfusion. Can. Med. Assoc. J. **156**, S9–S26 (1997)
12. Busch, M., et al.: A new strategy for estimating risks of transfusion-transmitted viral infections based on rates of detection of recently infected donors. Transfusion **45**, 254–264 (2005)
13. Carson, J., et al.: Red blood cell transfusion: a clinical practice guideline from the AABB. Ann. Int. Med. **157**, 49–58 (2012)
14. Goodnough, L.: Transfusion medicine: looking to the future. Lancet **361**, 161–169 (2003)
15. http://thebloodytruth.com/risky-business-the-relative-infectious-risks-of-blood-transfusion (2012)
16. Li, G., et al.: Incidence and transfusion risk factors for transfusion-associated circulatory overload among medical intensive care unit patients. Transfusion **51**(2), 338–343 (2005)
17. Infectious risks of blood transfusion. Blood Bull. **4**(2) (2001). America's Blood Centers, Washington, DC
18. Non-infectious serious hazards of transfusion. Blood Bull. (2002). America's Blood Centers, Washington, DC
19. Toy, P., et al.: Transfusion-related acute lung injury: incidence and risk factors. Blood **119**(7), 1757–1767 (2012)
20. US General Accounting Office. Blood Supply: Transfusion-associated risks. GAO/PEMD-971. US Government Printing Office, Washington, DC (1997)
21. Sandler, G.: Transfusion reactions treatment and management. Medscape (2016)
22. Schwartz, S., Blumenthal, S.: Exogenous hemochromatosis resulting from blood transfusion. Blood **3**, 617–640 (1948)
23. Mir, M.: Transfusion-induced iron overload. Medscape (2016)
24. Holnessa, I., Knippena, M., Simmonsa, L., Lachenbrucha, P.: Fatalities caused by TRALI. Transfus. Med. Rev. **18**(3), 184–188 (2004)
25. Bux, J., Sachs, U.: The pathogenesis of transfusion-related acute lung injury (TRALI). Br. J. Haematol. **136**(6), 788–799 (2007)

26. Choat, J., Maitta, R., Tormey, C., Wu, Y., Snyder, E.: Transfusion reactions to blood and cell therapy products. In: Hoffman, R., Benz Jr., E.J., Silberstein, L.E., Heslop, H.E., Weitz, J.I., Anastasi, J. (eds.) Hematology: Basic Principles and Practice, 6th edn. Elsevier Saunders, Philadelphia (2013)

27. Goodnough, L.: Transfusion medicine. In: Goldman, L., Schafer, A.I. (eds.) Goldman-Cecil Medicine, 25th edn. Elsevier Saunders, Philadelphia (2016)

28. Boonyasampant, M., Weitz, I., Kay, B., Boonchalermvichian, C., Leibman, H., Shulman, I.: Life-threatening delayed hyperhemolytic transfusion reaction in a patient with sickle cell disease: effective treatment with eculizumab followed by rituximab. Tranfusion 55(10), 2398–2403 (2015)

29. Pirenne, F., Narbey, D., Chadebech, P., Mekontso-Dessap, A., Bartolucci, P., Rieux, C., Habibi, A.: Incidence and risk of delayed hemolytic transfusion reaction in sickle cell disease patients based on a prospective study (2016)

30. Sahu, S., Hemlata, A.: Adverse events related to blood transfusion. Indian J. Anaesth. 58(5), 543 (2014)

31. Perrotta, P., Snyder, E.: Non-infectious complications of transfusion therapy. Blood Rev. 15(2), 69–83 (2001)

32. Walker, R.: Special report: transfusion risks. Am. J. Clin. Pathol. 88(3), 374–378 (1987)

33. Perkins, H., Payne, R., Ferguson, J., Wood, M.: Nonhemolytic febrile transfusion reactions. Vox sanguinis 11(5), 578–600 (1966)

34. Domen, R.: Adverse reactions associated with autologous blood transfusion: evaluation and incidence at a large academic hospital. Transfusion 38(3), 296–300 (1998)

35. Menis, M., Anderson, S.A., Forshee, R., Mckean, S., Gondalia, R., Warnock, R., Izurieta, H.: Febrile non-hemolytic transfusion reaction (fnhtr) occurrence among inpatient elderly medicare beneficiaries as recorded in large administrative databases in 2011. Transfusion 53, 140A (2013)

36. Waters, A.H.: Post-transfusion purpura. Blood Rev. 3(2), 83–87 (1989)

37. Yokoyama, A., Dezan, M., Costa, T., Aravechia, M., Mota, M., Kutner, J.: Diagnosis and management of post-transfusion purpura-case report. Blood 122(21), 4834 (2013)

38. Rohde, J., et al.: Health care–associated infection after red blood cell transfusion: a systematic review and meta-analysis. JAMA 311(13), 1317–1326 (2014)

39. Cázares-Benito, M., Cázares-Tamez, R., Chávez, F., Díaz-Olachea, C., Ramos-García, A., Díaz-Chuc, E., Lee-González, B.: Impact on costs related to inadequate indication of blood transfusion. Medicina Universitaria (2016)

40. Husted, H., Holm, G., Jacobsen, S.: Predictors of length of stay and patient satisfaction after hip and knee replacement surgery: fast-track experience in 712 patients. Acta Orthop. 79, 168–173 (2008)

41. Pulido, L., Ghanem, E., Joshi, A., Purtil, J., Parvizi, J.: Periprosthetic joint infection: the incidence, timing, and predisposing factors. Clin. Orthop. Relat. Res. 466, 1710–1715 (2008)

42. Díaz, M., Borobia, A., Erce, J., Maroun-Eid, C., Fabra, S., Carcas, A., Muñoz, M.: Appropriate use of red blood cell transfusion in emergency departments: a study in five emergency departments. Blood Transfus. 15(3), 199 (2017)

43. Brandis, K., Richards, B., Ghent, A., Weinstein, S.: A strategy to reduce inappropriate red blood cell transfusion. Med. J. Aust. 160(11), 721–722 (1994)

44. Corwin, H., et al.: The CRIT study: anemia and blood transfusion in the critically ill—current clinical practice in the United States. Crit. Care Med. 32(1), 39–52 (2004)

45. Vamvakas, E., Blajchman, M.: Transfusion-related mortality: the ongoing risks of allogeneic blood transfusion and the available strategies for their prevention. Blood 113(15), 3406–3417 (2009)

46. Sihler, K., Napolitano, L.: Complications of massive transfusion. Chest J. **137**(1), 209–220 (2010)
47. Koch, C., et al.: Morbidity and mortality risk associated with red blood cell and blood-component transfusion in isolated coronary artery bypass grafting. Crit. Care Med. **34**(6), 1608–1616 (2006)
48. Facts About Blood and Blood Donation. https://www.bnl.gov/hr/blooddrive/56facts.asp
49. Liumbruno, G., Bennardello, F., Lattanzio, A., Piccoli, P., Rossetti, G.: Recommendations for the transfusion of red blood cells. Blood Transfus. **7**(1), 49 (2009)
50. Taylor, R., et al.: Red blood cell transfusions and nosocomial infections in critically ill patients. Crit. Care Med. **34**(9), 2302–2308 (2006)
51. Adams, R., Lundy, J.: Anesthesia in cases of poor surgical risk some suggestions for decreasing the risk. J. Am. Soc. Anesthesiol. **3**(5), 603–607 (1942)
52. Options/Alternatives to Blood Transfusions. http://bloodless.com.br/optionsalternatives-to-blood-transfusions/
53. Fradkin, D., Muchnik, I.: Support vector machines for classification. In: DIMACS Series in Discrete Mathematics and Theoretical Computer Science, vol. 70, pp. 13–20 (2006)
54. Suykens, J., Vandewalle, J.: Least squares support vector machine classifiers. Neural Process. Lett. **9**(3), 293–300 (1999)
55. Ho, T. K.: Random decision forests. In: Proceedings of the Third International Conference on Document Analysis and Recognition, vol. 1, pp. 278–282 (1995)

Author Index

Printed in the United States
By Bookmasters